Cost Engineering Systems for Manufacturing Enterprises

Xiaoyi Liu

Cost Engineering Systems for Manufacturing Enterprises

Methods, Processes and Experience Reports from Practice

中国工信出版集团
電子工業出版社
PUBLISHING HOUSE OF ELECTRONICS INDUSTRY
http://www.phei.com.cn
Springer

Xiaoyi Liu
Chengdu, Sichuan, China

ISBN 978-981-96-5233-4 ISBN 978-981-96-5234-1 (eBook)
https://doi.org/10.1007/978-981-96-5234-1

Jointly published with Publishing House of Electronics Industry
The print edition is not for sale in China (Mainland). Customers from China (Mainland) please order the print book from: Publishing House of Electronics Industry.

This Springer imprint is published by the registered company Springer Nature Singapore Pte Ltd.
The registered company address is: 152 Beach Road, #21-01/04 Gateway East, Singapore 189721, Singapore

Preface I: Current Bottlenecks and Future Prospects for Manufacturing Companies

The current economic environment facing manufacturing companies:

- Rising costs for labor, materials and energy.
- The market and customers are demanding more and more intelligent and connected products, and manufacturers need to invest more in research and development to maintain their market share and attract customers.
- As the Chinese market continues to open up, falling product prices will further reduce the profitability of manufacturing companies amid increasingly fierce competition.
- Uncertainty about the future of the global economy, trade conflicts making consumers and investors more cautious.

On August 8, 2016, the State Council issued a notice on the work program to reduce enterprise costs in the real economy ([2016] No. 48).

In addition to the State's efforts to further reduce the costs of manufacturing enterprises at the policy and tax levels, this work program on reducing the costs of enterprises in the real economy also provides specific guidance on how manufacturing enterprises can better realize their own potential to reduce costs and improve efficiency, excerpts from the work program, e.g.:

- "guide companies to manage innovation and lean production and to use information technology tools to reduce costs. Encourage enterprises to make full use of the new generation of information technology and other means to improve internal management, to innovate marketing models and to increase efficiency".
- "Strengthen the promotion of advanced technologies and encourage enterprises to improve target cost management. Improve measures to promote and support enterprise restructuring and technological innovation, and support the promotion of various technologies that can effectively reduce enterprise costs to encourage enterprises to continuously improve production efficiency."

In many manufacturing enterprises in China, the traditional cost management model is still applied in terms of costs, which has the following disadvantages:

(1) Although the company management (including the boss) declares that cost is important, it is difficult to find an organizational structure dedicated to cost in the company, only under the finance department or under the purchasing department or even under the project management department we can find some "cost management" staff, the company does not treat cost engineering as an independent functional department, similar to R&D or quality departments, let alone have adequate staffing. A large manufacturing company with 50,000 employees has several thousands of R&D employees, but only 25 employees for "cost management", so it is impossible to fully promote and achieve cost reduction and efficiency in a practical way.
(2) Most manufacturing companies lack a systematic and comprehensive cost engineering approach and do not have a cost engineering process embedded in the entire product process; some companies do carry out cost reduction and efficiency improvement work (VAVE: Value Analysis and Value Engineering), but only sporadically, especially only then due to the economic downturn and sharp decline in sales, and are forced to carry out cost optimization, resulting in cost reduction and efficiency improvement not being part of the company's daily business.
(3) R&D engineers, due to their technical background or lack of experience, tend to think that it is the highest "art of engineering" to realize a function and develop a component that works reliably and that it is not their job to achieve this at the cheaper price. In fact, they believe that cost is only the responsibility of the "cost management" team. Manufacturing companies need a performance measurement mechanism (KPI) to achieve target costs that involves all employees and that really needs to be implemented. Is it motivating to talk about reducing costs and increasing profits for the company without involving employees in the success with an incentive?

Manufacturing companies are under enormous (cost) pressure and some are even in a situation where it is a matter of survival. There are always winners and losers in the market. What is the right way out for manufacturing companies?

One important way is "cost engineering". The reason why "engineering" is added after the word "cost" is that it has a profound meaning here. 85% or more of product costs are "designed", or "engineered", not "managed".

Cost engineering begins with product definition and development, optimizing product design, selecting the most appropriate materials and manufacturing processes, managing engineering changes, optimizing logistics costs, etc., to reduce product costs while ensuring product function and quality.

Cost engineering is creative engineering design work at a higher level, so to speak, looking for optimization solutions for existing engineering designs rather than just calculating and summarizing cost data. It has led many managers to believe that costs only need to be "managed", leaving a large scope for optimizing costs "in engineering" untapped.

Now is the time to change this mindset!

The author will use the term "cost engineering" throughout the book.

The development of cost engineering began in 1948 when L. D. Miles, an engineer at GE, first introduced the idea of matching the function and value of a product, then known as value management.

Since the 1990s, some internationally advanced manufacturing companies have gradually realized that cost engineering can be of great help to enterprises, have begun to separate cost teams from individual functional departments, such as cost accounting in the finance department, quotation evaluation in the procurement department, and lean production in the production lines, to form an independent cost engineering department for the entire product process within the organizational framework of enterprises, and have gradually established a complete cost engineering system with related processes and methods.

To date, some large manufacturing companies in Europe, the USA and Japan have established a cost engineering department, some of which are called cost engineering, others are called value analysis and value engineering (VAVE), and Japanese companies are called original price planning (original price is the Japanese use of Chinese characters, and means "cost").

In an environment of increasingly fierce competition and economic and trade conflicts, it is particularly important and urgent to introduce cost engineering to manufacturing companies in order to improve their global competitiveness and increase their profitability.

Based on 25 years of practical experience and methodological research in German and Chinese automotive companies, the author presents in this book for the first time a systematic introduction to the cost engineering system, cost engineering methods and their practical application in the entire product process and provides the reader with a variety of practical examples for cost optimization of components and products.

The book begins with the origins of cost engineering and describes in detail the entire product process as well as the tasks and responsibilities of cost engineering in the product process. In the introduction to the methods of cost engineering, the various methods of cost engineering are presented systematically and comprehensively, based on decades of practical work experience in manufacturing companies and theoretical research on methods of cost engineering, accompanied by numerous practical cases.

How can these cost engineering methods be applied in the daily work of a company and how can a company benefit from the advantages of cost engineering?

First of all, manufacturing companies urgently need a plug and play cost engineering model that can be used immediately, from the setting of objectives to the application of methods and the management process. The author shows practical examples of plug and play cost engineering models for manufacturing companies based on his work at Mercedes-Benz in Germany and Great Wall Motor in China.

To reach the next level in cost engineering, manufacturing companies need to establish a cost engineering system in the company's organizational structure and define the process of cost engineering in the product process; building a fully established cost engineering system is a necessary way for manufacturing companies to gain a foothold in the market and develop sustainably in the future.

Another mechanism is the modular strategy. Based on the experience of developing the modular strategy in the Daimler Group in Germany, the author systematically describes the methods, processes and examples of the modular strategy.

In the face of the rapid development of digitalization and artificial intelligence, the author proposes the concept of digitalization of cost engineering, which is realized through cost database, cost calculation and analysis software, and AI-assisted design, and the digitalization of cost engineering will gradually lead product design to an intelligent stage.

At the system level, manufacturing enterprises not only need quality and production systems, but also a complete cost engineering system to ensure the demand for high value-added jobs and employment and improve the international competitiveness of enterprises.

Looking to the future:

Like the quality system standard (ISO), the cost engineering system also needs a standard, and companies that do not meet the cost engineering system standard will lack the cost engineering certificate, just like a quality system ISO 9001, and therefore lose the trust of customers and orders because their own companies do not professionalize cost engineering, do not implement it effectively, and are not good enough on cost efficiency.

Cost databases, cost calculation and analysis software, AI-supported cost engineering will be important tools for manufacturing companies to tap further cost optimization potential.

Chengdu, China Dr. Xiaoyi Liu
May 2020

Preface II: By the Author on the Publication of the English Edition

Since the Chinese edition of Cost Engineering Systems for Manufacturing Enterprises was published by Publishing House of Electronic Industry (PHEI) on August 1, 2020, and the German edition was jointly published by Springer Verlag and PHEI on September 29, 2022, it has been very well received by managers and technicians in Chinese and German-speaking manufacturing enterprises, has been applied in their daily work of cost reduction and efficiency improvement, and is now the English edition also jointly published by Springer Verlag and PHEI. The author expects that readers in English-speaking regions or those who can read English books will be able to read this book and apply it to the daily optimization of product design, product manufacturing, and product sales and after-sales service, in order to reduce costs and increase efficiency, improve profitability, and enhance market competitiveness in such an "introspective" and competitive market.

Chengdu, China
January 2025

Dr. Xiaoyi Liu

Testimonial by Konghui Guo, Member of the Chinese Academy of Engineering

Since the reform and opening-up, China's manufacturing industry has developed rapidly and vigorously over the past 40 years. China has become the country with the world's most comprehensive range of industrial categories, with technology-intensive products and services such as PCs, smartphones, the Internet, new energy vehicles, and cashless mobile payments entering the homes of ordinary consumers, and CNC machine tools, industrial robots and automated production systems being used in many manufacturing enterprises.

Chinese manufacturing companies are catching up with the advanced companies in Europe, North America and Japan. To keep pace with these advanced companies or even surpass them in the near future, Chinese manufacturing companies have not only to innovate their products but also to improve their cost efficiency to survive the fierce international competition.

The function and quality of a product (customer characteristics) must be harmonized with the costs. Whoever can achieve the best cost under the premise of fulfilling the function and quality is the winner, and whoever can provide customers with the best product value will win more customers and increase the company's profitability and market competitiveness.

How can we improve the cost efficiency of our products and reduce product costs?

First, a manufacturing company needs a cost engineering team within the company's organizational framework.

Secondly, it needs the systematic application of numerous cost engineering methods in the product development and manufacturing process, such as reference cost calculation, value analysis and value engineering (VAVE) of products, and Design to cost, which are embedded in the entire product process.

Finally, the company needs a performance assessment mechanism (KPI) with all stakeholders to achieve the product cost targets.

Product costs are engineered, and 85% of product costs are determined in the early design phase. Optimizing the functional design of products, selecting the right materials and applying advanced manufacturing processes are the key starting points for optimizing product costs.

Based on 25 years of practical experience and methodological research at Mercedes-Benz in Germany and Great Wall Motor in China, the author, Dr. Xiaoyi Liu, presents for the first time in China a systematic introduction to the systems, methods and related processes of cost engineering embedded in product development, product manufacturing and the entire product life cycle process in advanced manufacturing enterprises.

Starting from the product process, this book *Cost Engineering Systems for Manufacturing Enterprises* introduces the cost engineering methods in detail, and each method description is followed by practical examples for the reader to understand and learn from.

I would like to take this opportunity to recommend this book to executives, product developers, financial analysts, and procurement specialists of manufacturing companies as a reference for decision-making to build a fully integrated cost engineering system, as a tool for design to target cost, and as an application and reference for cost analysis methods and business negotiation skills.

Looking ahead, manufacturing companies will definitely enter the digitalized and AI-assisted era, and with the modularization of products and the intelligence of manufacturing, cost engineering will also enter the digital and AI-assisted realm. Just as computer-aided engineering (CAE) creates a leap forward for engineers in optimizing product design, digitized cost engineering will enable engineers to take the next leap forward in optimizing product costs.

May 2020

Konghui Guo
Member of the Chinese Academy of Engineering
Jilin University
Changchun, China

Testimonial from Guoxin Zhang, General Manager of the Haval Subsidiary of Great Wall Motor Co. in Tianjin

China's economy is in a transitional phase where the focus is no longer on size and quantity, but on quality and efficiency. At the beginning of reform and opening up, we were coming out of an era of material scarcity, and economic development was mainly to satisfy consumer demand for "having". From nothing to something, from something to better, from better to that which fulfills customers' individual needs, the entire market has changed fundamentally. In the early days, the domestic market was undersupplied, and consumers with limited purchasing power, while at the same time consumer enthusiasm was high, and gradually developed into an era of supply and demand, in which enterprises could produce on a large scale and the products produced were easily accepted by consumers.

With the development of the times, major consumer groups are constantly changing, tending to individualization and diversification, while the supply capacity of the industry is rapidly increasing, and competition in all fields has become much fiercer with brand, price and innovation. People's perception of cost has also changed.

Dr. Liu's efforts to reconstruct everyone's cost perception during this period meet the needs of these changing times.

For a company's products to stand out in the competition, fundamentally meet customers' needs and be the best in the market, they must not only exceed customers' expectations, but also meet customers' expectations in terms of price, whether it is good value for money or exceeding good value for money, all this requires accurate cost planning and execution.

The key to the cost of a product is in the planning and then in the execution. Now we say that there is a lot of room for improvement, but the room for improvement will always be to hit the ceiling. Regardless of the industry, a company's operational vision cannot be separated from this formula: profit = sales price – cost.

The changing customer needs, the demand for high quality, the demand for low prices, the demand for personalized products are undoubtedly gradually eating away at the core of a company's survival (profit). The general cost elements such as labor costs, capital costs, land costs, raw material costs, energy costs, environmental costs, logistics costs, administrative costs are increasing year by year.

The proportion of China's working population has fallen to 71.8% in the past seven years and is expected to fall to 70% by 2023, and the average wage of employees in urban areas in China reached CNY 88,300 per year in 2019, an increase of 2.42 times from 2010, so China is gradually losing its labor cost advantage. These have already had a significant impact on the sustainability of China's manufacturing industry. As the profit margin is gradually disappearing, manufacturing enterprises need to conduct a comprehensive study on cost planning and control if they want to maintain their original profit margins.

Now the post-90s and post-00s have gradually become the main body of consumers, this group has an extravagant personality and has "enough money" from birth, hoping for products with a sense of technology, high value and stylish styling, which poses a new challenge to our traditional manufacturing industry. This trend is gradually intensifying, and we manufacturers need to shift from B2C to B2M, which puts higher demands on the speed of product construction and product delivery. Profitability determines the survival of a company. With sufficient profitability, we can continue to innovate, invest more in research and development, purchase more equipment and attract more talents, so that we can achieve sustainable business.

Cost management is a systematic project ranging from market knowledge and analysis, product planning and development, production control to customer service, taking into account every link of cost engineering to plan costs on the spot to exceed customer expectations. Our original focus was more on the management of a specific link, but cost management is also a chain management, and to be efficient, each link must be improved simultaneously. There are many links, and the cross-fertilization of technology and management, which require an overall framework.

The product design stage is the optimal period for cost engineering, to meet the market demand under the premise of reasonable cost to develop the product, take the automobile as an example, these products are more than 2000 assemblies, more than 30,000 components, with thousands of direct suppliers, how to choose the right supply chain in the world, how to solve the design excellence, cost control, a number of systems must be used to support.

For the production chain, such as the energy consumption of a factory, many companies only carry out periodic result statistics. Is the energy consumption appropriate? Is there room for optimization? Since there are no statistics on the smallest unit of energy consumption, it is only possible to meet "running, bubbling, dripping and leaking" monitoring and control requirements from a management perspective, but there is no way to achieve the optimal utilization of energy combinations and energy consumption of energy-consuming equipment. How does this phenomenon occur? The process also requires a large number of employees to read the meter readings and then perform a summarized statistical analysis. The punctuality does not meet the company's cost control requirements. Analyzing the energy structure is still not enough because China is a huge country and the prices of different energy sources vary from place to place, so accurate cost planning is required for reasonable reconciliation.

Meanwhile, there are many new technologies and methods such as Big Data, Artificial Intelligence (AI) and the Internet of Things, which open up new ways and opportunities for cost planning and cost management. These innovations offer new tools for cost planning and execution, and these tools need to have an overall framework to play a role of $1 + 1 > 2$. The transformation of manufacturing companies is a comprehensive, three-dimensional process. At this time when the COVID19 pandemic is ravaging the world, when we are discussing the survival and development of enterprises, when we are talking about good management, when everyone is discussing cost and competitiveness, and when innovative ideas require innovative methods, so when we hear that Dr. Liu will talk about cost engineering and give management suggestions from theoretical and practical frameworks, we are always looking forward to showing more people that cost engineering must be integrated into the management bloodline of every manufacturing enterprise. We also hope that more and more people will realize that cost engineering can be integrated into the management of every manufacturing enterprise, so that the organization can be more efficient and faster, and the competitiveness of the enterprise will be improved.

Tianjin, China
May 2020

Guoxin Zhang

Testimonial from Dr. Xianqiang Lu, Chief Engineer at the Product Development Center of Chery Automobile Co.

I had the pleasure of reading the book *Cost Engineering System for Manufacturing Enterprises* by Dr. Xiaoyi Liu. Based on his decades of cost engineering experience in domestic and international companies, Dr. Liu systematically explains what cost engineering is and how to use cost engineering processes and tools to perform effective cost optimization throughout the product life cycle. This is the best book on cost management I have read in China. I would like to introduce this book to my colleagues in the industry and I am convinced that it will be of great help to you in your future work.

Wuhu, China
May 2020

Dr. Xianqiang Lu

Introduction

After living in Germany for 30 years, I returned to China in the fall of 2013. In my spare time, especially in the past 3 years of entrepreneurship, I had written dozens of articles on cost engineering methods and practical examples from industry, most of which were published in Liu Xiaoyi's column of Gasgoo-Auto.

In the spring of 2019, I was fortunate to meet Dr. Tingru Li from the Talent Exchange Center of the Ministry of Industry and Information Technology (MIIT) at an appointment ceremony of the Think-Tank Expert for the New Energy and Intelligent Connected Automotive Industry, and upon her encouragement, I decided to write this monograph on the cost engineering system for manufacturing enterprises, which records my industry experience in Germany and in China. The book documents my experiences from cost engineering practice and research on cost engineering methods at Mercedes-Benz and Great Wall Motor.

Everything is done from the perspective of the company and the practice, and the reader will feel like they have experienced it themselves. If the reader applies the methods in the book or uses the examples to optimize product costs, make production lines more efficient and increase the company's profit, this is the result the author is aiming for.

I would like to express my sincere thanks to Ms. Xiaohong Miu, editor of Publishing House for Electronic Industry, for her help and support in publishing this book, her suggestions for improving the outline and text, and her proofreading.

Shanghai, China
May 2020

Dr. Xiaoyi Liu

Contents

1 The Different Types of Systems in Manufacturing Enterprises 1
1.1 Why Most Manufacturing Companies Do Not Have a Cost Engineering System 2
1.2 The Cost Engineering System 3
1.2.1 Positioning of Cost Engineering in the Manufacturing Company 3
1.2.2 The Practical Application of Cost Engineering Methods in the Product Process 5
1.2.3 Cost Engineering—Embedded in the Product Process 6
1.2.4 R&D Cost Planning 7
1.2.5 Tool Cost Planning 10
1.3 The Cost Engineering System Standard (ISO/GB/DIN) 12

2 The Practice of Cost Engineering in Manufacturing Companies at Home and Abroad 15
2.1 Cost Engineering in the Practice of European Manufacturing Companies 17
2.2 Original Price Planning in Practice in Japanese Manufacturing Companies 19
2.3 The Practice of Cost Estimators in American Manufacturing Companies 20
2.4 Cost Management in the Practice of Chinese Manufacturing Companies 21

3 Product Function and Product Value 23
3.1 The Origins of Value Management 23
3.2 The Relationship Between Product Function/Performance and Product Value 23
3.3 The Essence of Value Management 27

4 The Product Process ... 29
4.1 Definition of the Product Process ... 29
4.2 Responsibility Matrix for the Product Process (RASIC) ... 41
4.2.1 Tasks and Responsibilities of R&D in the Product Process ... 41
4.2.2 Role and Responsibility of Quality in the Product Process ... 41
4.2.3 Tasks and Responsibilities of Cost Engineering in the Product Process ... 43

5 Cost Engineering Methods ... 45
5.1 Multi-dimensional ABC Analysis ... 45
5.2 Reference Cost Calculation ... 47
5.3 Design to Cost (Design to Cost) ... 53
5.3.1 Morphological Box ... 54
5.3.2 Function-Cost Matrix ... 56
5.3.3 Benchmark Function ... 59
5.4 Value Analysis and Value Engineering (VAVE) of Components and Assemblies ... 60
5.4.1 Value Analysis ... 61
5.4.2 Value Engineering ... 68
5.5 Linear Regression Analysis ... 68
5.6 Best of Bench ... 71
5.7 Portfolio Analysis Method ... 71
5.8 Analytical Functional Analysis of Manufacturing Costs and Production Volume ... 72
5.9 Business Negotiation Strategy and Tactics ... 74
5.9.1 Negotiation Goals and Motivations ... 75
5.9.2 Preparing for the Negotiation ... 75
5.9.3 The Tactics of the Negotiation Process and the Art of Negotiation ... 76
5.9.4 Concluding Negotiations and Summaries ... 78
5.10 Business Case Calculation (Net Present Value: NPV) ... 82

6 How to Apply Cost Engineering Methods to Reduce Component Costs ... 85
6.1 Value Analysis and Value Engineering (VAVE) Guided by Standardization Questions ... 85
6.1.1 The Methodology of the VAVE Workshop ... 86
6.1.2 Practical Examples of VAVE ... 87
6.2 Avoiding Over-Engineering ... 93
6.2.1 Product or Component Specification ... 95
6.2.2 Performance Requirements that Go Beyond the Customer's Normal Area of Application ... 98

6.2.3 Component Lifetime Beyond the Product Lifetime 99
6.2.4 Practical Examples to Avoid Over-Engineering 100
6.3 Cost Engineering and Lightweight Construction 101
6.3.1 Cost-Weight Function 102
6.3.2 Practical Examples of Cost and Weight Reduction 102

7 The Target Cost Process of a Product 107
7.1 The First Customer 111
7.2 Determining Target Product Costs 112
7.2.1 Determining the Target Costs 112
7.2.2 Subdivision and Adjustment of Target Costs 115
7.2.3 The Target Cost Achievement Path 117

8 Optimization of Manufacturing Costs 121
8.1 Toyota Production System (TPS) 121
8.2 Mercedes-Benz Production System (MPS) 123
8.2.1 Value Stream Mapping 123
8.2.2 Method of Data Collection at the Workplace 127
8.2.3 7V (7 Types of Waste to Avoid) 127
8.2.4 5S (5S for a Good Workplace) 129
8.2.5 3M (Avoid 3M in Production) 130

9 Modular Strategy: Cost Engineering 2.0 133
9.1 Why Do Manufacturing Companies Need a Modular Strategy? 133
9.2 How to Implement Cost Engineering 2.0? 134
9.2.1 Analysis Phase 137
9.2.2 Technical Design Phase 138
9.2.3 Implementation Phase 141

10 Cost Engineering for New Energy Vehicles 143
10.1 The Key Customer Benefits of New Energy Vehicles 143
10.2 Cost Engineering in Practice for New Energy Vehicles 144
10.2.1 Business Case Calculation (NPV) and Sensitivity Analysis 144
10.2.2 Calculation of the Reference Cost of Lithium Battery Cells 144
10.2.3 Comparison of the Structure and Costs of Directly and Indirectly Driven Electric Motors 149

11 Digitalization of Cost Engineering 155
11.1 Digitalization of the Product Process 155
11.2 CAE for Optimizing Product Material Costs 156
11.3 CAE Optimization of Manufacturing Costs 156
11.4 Cost Database 158
11.5 Cost Software 159

Conclusion 163

References 165

Index 167

About the Author

Dr. Xiaoyi Liu was born in Ya'an, Sichuan Province, and successfully took the first national university entrance examination after the reform and opening up in 1977 to enter the Department of Electrical Engineering and Computer Science of Shanghai Jiaotong University. After receiving his doctorate degree from Technische Universität Berlin on "Contribution to the Simulation of Pneumatic and Electrostatic Painting Process" with honors (summa cum laude), he worked for Daimler AG (Mercedes-Benz) in Germany for 21 years, where he was responsible for the research of new production processes, the series development of powertrain electronic control systems, the entire vehicle planning of new energy vehicle and cost engineering, and led several major projects to reduce costs and increase efficiency (PKO, Optima, CTX, CORE, etc.). After returning to China, he worked in the upper management circle at Great Wall Motor's R&D center, where he was responsible for the cost engineering and lightweighting of vehicles of all brands and led the Group's cost reduction and efficiency improvement work.

Dr. Xiaoyi Liu has received several awards at home and abroad

- German Physicist Carl Ramsauer Prize.
- Environmental Award of the City of Ulm for the development of environmentally friendly production processes for the manufacture of integral aircraft components.
- Award for the best cost optimization and efficiency improvement of Great Wall Motor.

And he is

- Visiting Professor at the College of Automotive Studies, Tongji University.
- Enterprise Supervisor of Postgraduate Students in Electrical Engineering at the Department of Computer Science and Electrical Engineering, Shanghai Jiaotong University.
- Postgraduate Enterprise Supervisor of Energy and Power Engineering College, Nanjing University of Aeronautics and Astronautics.

- Member of the New Energy and Intelligent Connected Automotive Industry Think-Tank Expert of the Talent Exchange Center of the Ministry of Industry and Information Technology.
- Lecturer of Cost Engineering appointed by Talent Exchange Center of Ministry of Industry and Information Technology, China Automotive Industry Training Base, Golden Finance, Shanghai Anting Government, and Changzhou Municipal Government.
- Author of "Liu Xiaoyi's column" of Gasgoo-Auto.

Dr. Xiaoyi Liu is the founder and CEO of CostKey Solutions (Shanghai) Co. Ltd.

Chapter 1
The Different Types of Systems in Manufacturing Enterprises

According to the definition of scientific and technical terms, a system refers to a certain set of things or the same kind of things combined in a certain order and internal context to form a whole, a whole composed of various subsystems, running through two core elements, one is a set of various subsystems, the other is working in a certain order and internal context.

For example, an automotive company has a research and development system with several "subsystems", which are as follows.

- Component design and validation department
- Overall vehicle integration and evaluation department
- Simulation and Calculation Department
- Series and Life Cycle Management department
- etc.

The "organization and internal linking" of the various departments is governed by the following guidelines and regulations.

- The job descriptions and the responsibility matrix of the departments (RASIC)
- Product requirements (specification)
- Timetables for product development
- Guidelines for engineering changes
- etc.

These guidelines and regulations serve to regulate the relationships within and between departments, while also defining the relevant standards and requirements.

Of course, manufacturing companies also have corresponding quality and production systems in addition to their R&D systems.

X. Liu, *Cost Engineering Systems for Manufacturing Enterprises*,
https://doi.org/10.1007/978-981-96-5234-1_1

1.1 Why Most Manufacturing Companies Do Not Have a Cost Engineering System

Before answering this question, the author would like to share a personal experience from recent years. The author visited dozens of senior managers in manufacturing companies, and when asked "What three issues have you been most concerned with in your company recently", the answer was: "1. revenue, 2. costs, 3. quality".

To the question "If you think the second most important issue is cost, what has your company done about cost in the last six months?", a common response is "let me think, oh, sorry, I can't remember".

This is true for the many manufacturing companies where management emphasizes the importance of costs but does nothing or rarely does anything to optimize them.

The main reasons for this are as follows:

(1) Many managers believe that costs are managed, and when asked how costs can be optimized, many will ask a counter question, "You mean cost control?" On the contrary, costs are engineered and need to be optimized through creative engineering solutions. Many manufacturing companies still leave their cost work in the finance, project management or even procurement departments and think that cost optimization is only about cost control or cost management. Manufacturing companies have still not recognized the importance of cost engineering. The key lies in its very name: cost engineering, where the word "engineering" means "development or technical solution":

Cost engineering is the use of technical means to optimize product costs

(2) Some manufacturing companies believe that costs are important and do a lot to reduce costs and increase efficiency, but the results are not very effective due to a lack of systematic cost engineering methods and processes.

(3) Many manufacturing companies view cost reduction and efficiency improvement as projects with limited duration, and do not undertake cost reduction projects when they are doing well, but must be forced to undertake cost reduction projects when they are doing poorly, instead of treating cost reduction as a daily business, just like R&D, production, quality, etc.

(4) Although cost engineering has its place in the organizational structure of manufacturing companies and some basic methods and procedures of cost engineering are in place, there is no enforceable performance evaluation mechanism (KPI) for implementation and target achievement, and departments are not motivated to do cost engineering systematically.

1.2 The Cost Engineering System

The concept of a cost engineering system was first introduced by the author in August 2017 in a publication "Cost engineering system in the automotive industry—key to reducing costs and increasing efficiency" in Gasgoo-Auto [1]. The future direction of cost engineering in manufacturing companies is to form a complete cost engineering system, just like the R&D system, the production system or the quality system.

Like a quality system, a cost engineering system requires an independent functional department within the company's organizational framework, a set of standards and specifications, a systematic approach and clear processes, and a performance assessment mechanism (KPI) based on cost targets, among other things. Figure 1.1 shows the three key factors of a cost engineering system.

1.2.1 Positioning of Cost Engineering in the Manufacturing Company

The first area is the organizational structure:

The positioning of the cost engineering department in the organizational structure of the company is extremely important, for example, the cost engineering department of a German OEM is a separate department that works in parallel with R&D, purchasing and production (see Fig. 1.2).

The advantage of this concept is that the product team, consisting of R&D, cost, purchasing, quality and production, pursues three common goals: Product Function, Cost and Quality.

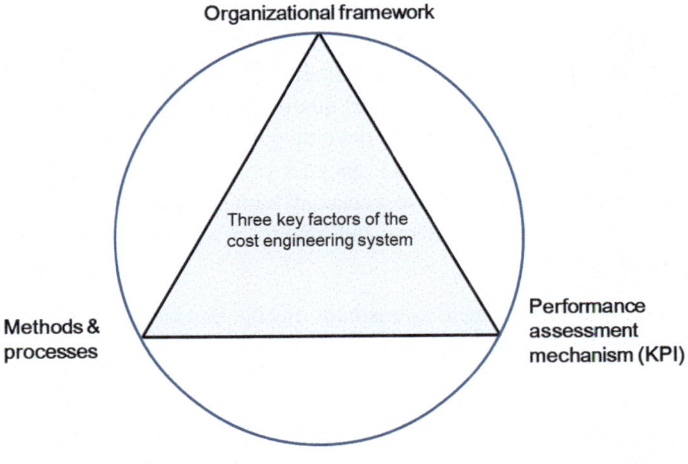

Fig. 1.1 Three factors of a cost engineering system

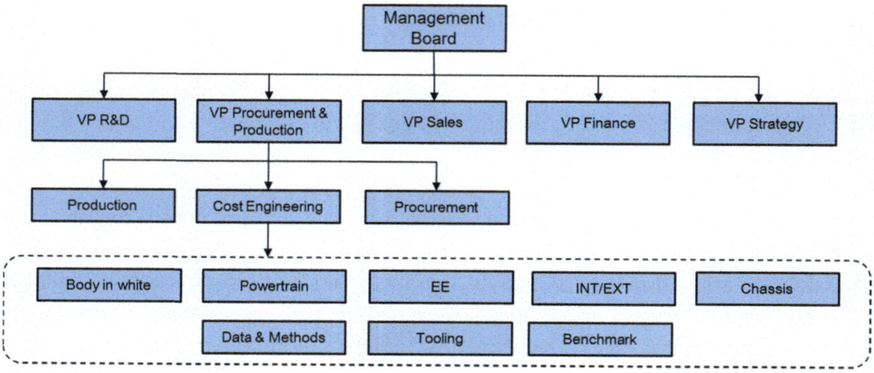

Fig. 1.2 Organizational structure of the cost engineering system

The R&D department is not solely responsible for product function, the cost engineering department is not solely responsible for costs, the purchasing department is not solely responsible for production supply, and the quality and production departments are not solely responsible for product quality and output, but for each product process there are defined inputs, methods, processes and outputs at the interfaces of these departments in order to achieve the overall optimization of product function, costs and quality.

in a manufacturing company.

The second area is methods and processes:

The cost engineering department is a technical department, and the cost engineers are usually experienced product development engineers and manufacturing process engineers who are familiar with customer requirements and manufacturing processes.

The cost engineer must be involved in product definition, design and manufacture from the outset, apply various cost engineering methods, propose technical optimization solutions and collaborate with colleagues from the research and development, purchasing, production and quality departments in order to achieve the best cost efficiency while ensuring the function and quality of the product.

The art of an engineer is not only to develop a product that meets the functional/performance requirements, but also to design a product that meets the functional/performance requirements at the lowest cost and with the required quality. Engineers here refer not only to R&D engineers, but also to cost, quality and manufacturing process engineers.

The third area is the performance assessment mechanism (KPI):

All departments of a manufacturing company involved in the product process must set appropriate cost target KPIs.

Cost engineering is a daily task for a manufacturing company, and it is not just about generating some VAVE ideas to reduce costs just because market pressure requires product cost reduction. Rather, design to cost takes place from day one of any new product definition, and engineering changes are subject to cost engineering management, as 85% of product costs are determined in the early stages of product definition and design. Of course, VAVE after series production is also essential.

In regions and countries such as Europe, North America and Japan, some manufacturing companies already have a cost engineering department, and some advanced companies have even established a complete cost engineering system, which effectively reduces costs while maintaining product function and quality for companies.

In the current volatile economic situation and tight market environment, manufacturing enterprises have an even more urgent need for a comprehensive cost engineering system to effectively reduce costs and increase efficiency, as well as improve the profitability and competitiveness of enterprises.

1.2.2 The Practical Application of Cost Engineering Methods in the Product Process

What are cost engineering methods? When it comes to VAVE (Value Analysis and Value Engineering), many people seem to know about it. Is this really the case?

The letters VAVE alone are divided into VA (Value Analysis) and VE (Value Engineering), with VE being used in the product development phase and VA in series production. Essentially, this is the value analysis of components and assemblies, one of many cost engineering methods.

There are dozens of cost engineering methods, which can essentially be divided into the following three categories, which the author will describe in more detail in later chapters.

Analysis category:

Reference cost calculation; value analysis of components and assemblies; multi-dimensional ABC analysis; market price and purchase volume analysis, etc.

Design to cost category (design at target cost):

Morphological box method; function-cost-matrix method; benchmark function method, etc.

Commercial category:

Business case calculation (NPV: Net Present Value); negotiation strategy and tactics, etc.

The product process includes the entire process of product definition, product development, product manufacturing, product sales and customer service and is the core process of a manufacturing company. The product process includes the elements of R&D, cost, quality and production. In other words: Cost engineering is primarily an important part of the product process.

Table 1.1 Product process and corresponding cost engineering methods

	Product process			
Application of cost engineering methods	Product definition	Product devlopment	Product manufacturing	Product sales, customer service
Business case calculation (NPV)	✓			
Design to cost		✓		
Reference cost calculation		✓		
Value engineering		✓		
Value analysis			✓	✓

Different cost engineering methods are generally used in the various phases of the product process.

Business case calculations (Net Present Value: NPV) are used in the product definition phase, where the NPV of the product over its entire life cycle is used as the basis for decisions on the development and production of a new product and as the basis for determining the target cost of the new product.

Value Engineering is used in the product development phase for the technical optimization of product solutions on the basis of design to cost and reference cost calculation.

Value analysis is used in the series production, sales and customer service phases to explore the potential for further cost optimization.

Table 1.1 describes the main tasks required for cost engineering throughout the product process (including product definition, product development, product manufacturing, product sales and after-sales service).

1.2.3 Cost Engineering—Embedded in the Product Process

Cost engineering is not an "optional" but a "mandatory" activity for manufacturing companies. Whether a manufacturing company considers cost engineering to be an essential part of its business depends solely on whether cost engineering is embedded in the company's overall product process.

As Fig. 1.3 shows, cost engineering is a parallel function to R&D, purchasing, production, quality and sales and customer service. Whether it is the development of a new product or the cost optimization of a series product, Cost Engineering is involved or takes the lead. In a production company, the departments for research and development, cost engineering, purchasing, production, quality and customer service form a matrix network with precisely defined inputs, methods, processes and outputs as well as responsibilities at each connecting node.

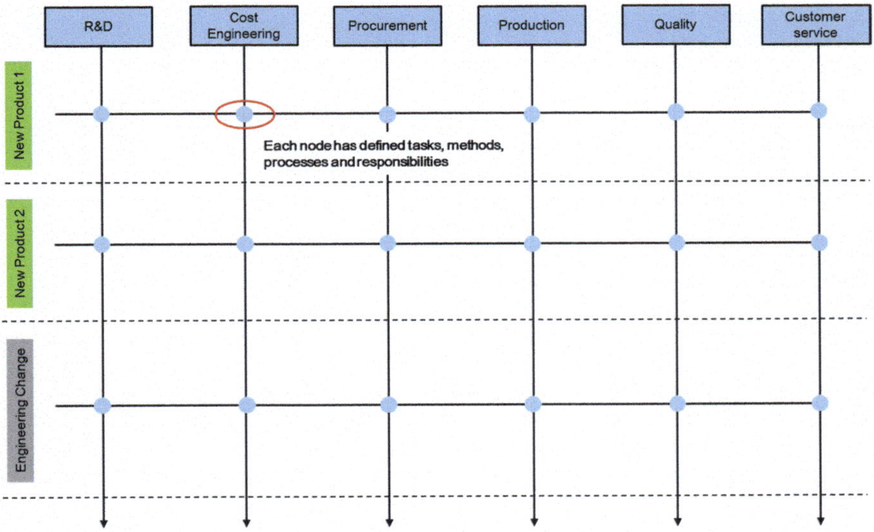

Fig. 1.3 Cost engineering in the product process

Looking at the development schedule of a complete vehicle (see Fig. 1.4), target costs and sales targets per vehicle must be defined at the (– 40) month node. At the (– 33) month node, after the preliminary styling work has been completed, a preliminary forecast of the costs for the complete vehicle is drawn up.

During the period (– 36 months from start to – 19 months), the designs for the main subsystems are produced at the target cost (design to cost) before the design data is frozen. During the period (– 19 months to 0 months), the Cost Engineering department must perform strict control and optimization for the engineering changes made during the development period.

In addition to reporting target cost achievement at (– 13)-months, a meeting will be held to confirm target cost achievement at (– 4)-months. After the (– 0)-month point (at the start of series production), the Cost Engineering department will take the lead in implementing continuous cost reduction and efficiency improvements (VAVE).

1.2.4 R&D Cost Planning

Section 1.2.3 describes the cost engineering embedded in the product process and its workflows with the aim of optimizing product costs. In addition, the development of a new product requires significant one-time investments in R&D costs, tooling costs and production facilities (including new and modified production facilities). Figure 1.5 illustrates the one-time investment required for the development and production of a new passenger car.

Fig. 1.4 Development schedule for a complete vehicle with cost milestones

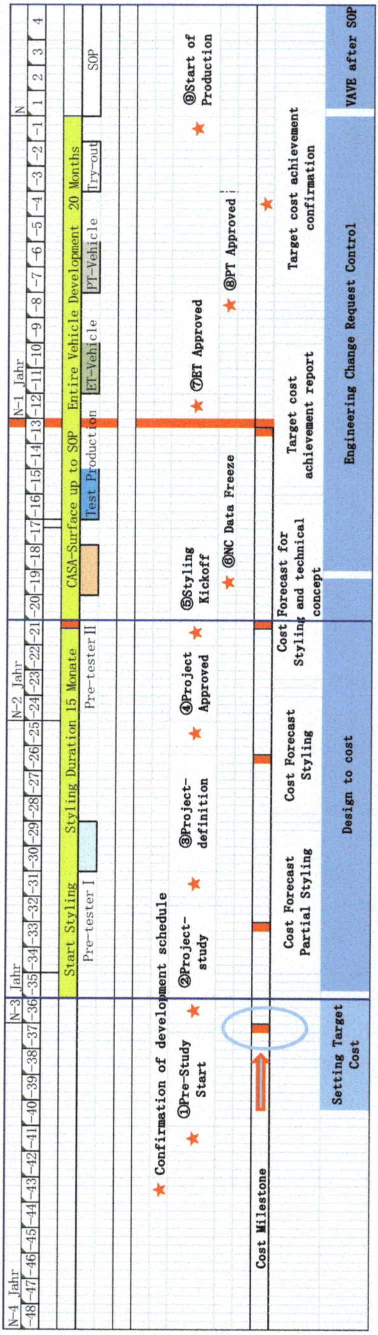

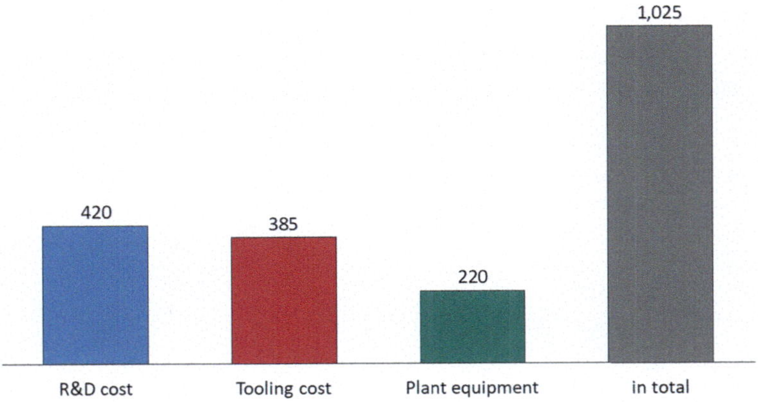

Fig. 1.5 One-time investments (R&D, tools, plant equiment) in million EUR

The total one-time costs for this new passenger car amount to €1.025 billion, of which €420 million is for research and development, €385 million for tools and €220 million for plant equiment.

The €420 million for research and development is a lot of money for an automotive company. How do you spend it carefully to maximize the ratio between output and input?

One of the main components of R&D costs is the cost of labor hours as well as the cost of prototypes (sample vehicles) and the cost of validation (Table 1.2).

The most important elements of R&D cost planning are Labor planning, prototype planning and validation planning. The aim of cost planning is to optimize the costs for working hours, prototypes and validation.

Firstly, we start with the working hours. Depending on the product, different labor hours are required. If a production company has not yet collected much experience data on labor hours, it needs to install a new software package—an IT system for recording labor hours.

For example, a German OEM in Germany has such a working-hours feedback system where each engineer involved in the development of a particular vehicle project has to log in weekly and record the corresponding working hours for that week for the vehicle project in question, its subsystems and the type of development activity.

At the end of the development of the vehicle project, the hours worked are fully itemized in the working hours feedback system, i.e. how many working hours were

Table 1.2 Cost accounting for research and development

Work (working hours)	Design, styling, CAE, etc.
Prototype (prototype vehicle)	Parts produced with "soft" molds (non-standard molds), 3D-printed parts
Validation of function and endurance tests	in the laboratory, on the test bench, in road tests

Table 1.3 R&D cost planning

	Current product	Next product in development: optimization potential
Working hours	Design	Part of the construction is outsourced to a third party
	Styling	Part of CAS modeling outsourced to third parties
	CAE	Increase in CAE and reduction of some validations
	Meetings	Reducing the number of meetings, creating a meeting structure within the company
Prototypes	Test bench test prototypes	
	Prototypes for road tests	Reuse
	Soft mold parts and 3D-printed parts	Time optimization, reduction in the need for parts made from soft molds and 3D printed parts
Validation	Subsystem validation	Increased use of CAE, reduction of subsystem validation
	Road tests	Reduction in the number of prototypes (samples) used, optimization of the combination of validation objects; optimization based on actual use by the customer
	Extreme use case test	Optimization based on the actual load spectra for customer use

spent on the design of each subsystem, how many working hours were spent on the validation of each subsystem, how many working hours were spent on the integration and validation of the whole vehicle; how many meetings were held (how many participants, how much time), etc.

A breakdown of the costs for prototypes and validation should also be provided.

Below you will find the author's analysis and recommendations for optimizing R&D costs based on his many years of experience in R&D cost planningTable 1.3.

1.2.5 Tool Cost Planning

Tooling cost planning is similar to R&D cost planning and requires a process of data accumulation. Figure 1.6 shows the tooling costs of 385 million EUR for a passenger car, which are distributed among the individual systems as follows.

Each of these systems consists of dozens of subsystems, and tooling cost planning consists of creating a tooling cost forecast for the new product at the subsystem level.

One or more tools are required for each subsystem, depending on the manufacturing process. The individual tools are calculated according to the reference cost model. Figure 1.7 shows a calculation example of an injection molding tool for the cover of a passenger car steering column switching module. The mold is designed for a horizontal injection molding machine with 5000 kN, with one cavity and a

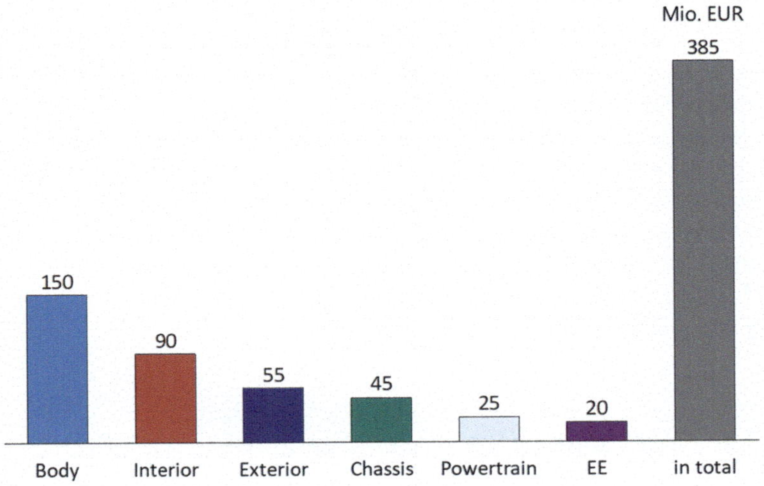

Fig. 1.6 Distribution of tooling costs by system for a passenger car

production cycle of 60 s. The dimensions of the mold are 446, 446 and 348 mm in length, width and height, and the weight of the mold is 544 kg.

The production of the tool is essentially divided into three phases:

- The first phase is the design of the tool and CNC programming,
- the second phase is the machining of the tool and

Quotation Calculation	**Supplier**			**Customer**		
Stem Form/Standard Part/Material			€			10,741 €
Heating Channel/Heating Nozzle			€			1,520 €
Surface (Graining/Coating)	h		€	0 h		67 €
Tooling Design/CAD	h	€/h	€	152,4 h	60 €/h	9,147 €
NC-Programming	h	€/h	€	78,6 h	60 €/h	4,713 €
Milling	h	€/h	€	7,7 h	55 €/h	426 €
Head Milling/CNC/HSC	h	€/h	€	325,4 h	61 €/h	19,848 €
Grinding	h	€/h	€	67,3 h	45 €/h	3,028 €
Turning	h	€/h	€	5,9 h	45 €/h	264 €
Eroding/Electrode/Wiring	h	€/h	€	344,3 h	53 €/h	18,246 €
Assembly/Handwork	h	€/h	€	243.4 h	45 €/h	10,954 €
Deep Hole Drilling	h	€/h	€	33,1 h	50 €/h	1,656 €
	h	€/h	€	h	€/h	€
Total	h	€/h	0 €	h	€/h	68,282 €
Sample/Test Report			€			4,028 €
Total Tooling Cost 1.)			115,000 €			84,587 €
of which slide machining	h					
Additional Tooling Cost	%		0 €	0 %		0 €
Reason:						
Total Tooling Cost 2.)			115,000 €			84,587 €

Fig. 1.7 Tooling cost break-down for the cover of a passenger car cladding tube switching module

- the third phase is the commissioning of the tool.

Based on the cost per man-hour of labor and machine, the time required for each phase, the cost of each phase can be calculated.

The reference cost of the tool for the cover was calculated at €84,587. The ability to calculate the reference costs for the tool is the basis for planning the tool costs and optimizing them. For example, from the detailing of the tool cost (see Fig. 1.7), the EDM process has a long working time of about 344.3 h; is it possible to replace part of the EDM process with high speed cutting (HSC)? The reason for this is that the machining speed of HSC is much higher than that of EDM. In the case of this tool, the supplier provided a cost estimate of €115,000.

1.3 The Cost Engineering System Standard (ISO/GB/DIN)

The quality assurance system of a production company complies with the ISO 9001 standard and the corresponding evaluation mechanism.

For the cost engineering system, the following questions are examined first:

(1) Does a manufacturing company need a cost engineering system?
(2) What are the criteria for a complete cost engineering system? What conditions must manufacturing companies fulfill?
3) What kind of assessment mechanism is required to issue a Cost Engineering System Certificate of Conformity to a manufacturing company?

The answer to the first question is yes, a manufacturing company needs a cost engineering system. In Sect. 1.2, the author described that cost engineering can only be a system if it is.

- has an independent department within the company's organizational structure,
- uses the cost engineering methods and processes embedded in the product process, and
- has a performance assessment mechanism (KPI) for the implementation of target costs.

In the author's opinion, the criteria for a cost engineering system are as shown in Table 1.4.

The author proposes a cost engineering system standard for all manufacturing companies. As with the quality system standard, manufacturing companies need an independent third party to issue a certificate that complies with the cost engineering system standard.

A manufacturing company with a certificate (ISO/GB/DIN) that complies with the cost engineering system standard is also an indirect indication of the company's efforts, standards and efficiency in cost engineering to optimize product costs while ensuring product quality and functionality for the customer, providing a "brand of

Table 1.4 Criteria for a cost engineering system

Element	Feature
Cost engineering organizational structure	Independent functional department, parallel to R&D, quality, production, purchasing, sales and customer service departments
	1:10 to 1:20 staffing in relation to R&D
	Cost engineers have experience in product development and production processes and are highly qualified
Cost engineering methods and processes	Embedded in the product process (new development products, series products)
	Determination of product target costs, target costs broken down by subsystems
	Business case calculation of products (NPV)
	Design at the target costs (design to cost)
	Reference cost calculation
	Control of engineering changes
	Cost reduction and efficiency increase in series production (VAVE)
Performance assessment (KPI)	Cost targets (cost reduction targets): KPI percentage greater than or equal to 25% to the entire KPI and actual implementation
	Cost engineering method training

efficiency label" for the company. The cost engineering system standard must be reviewed at regular intervals (e.g. every two years).

Chapter 2
The Practice of Cost Engineering in Manufacturing Companies at Home and Abroad

Since the 1990s, some large international manufacturing companies have gradually realized that cost engineering can be of great benefit to enterprises, and have begun to withdraw teams from departments such as cost accounting in the finance department, quotation evaluation in the purchasing department, and lean production in the production line, etc., to form an independent cost engineering department within the company's organizational framework and throughout the product process, and this leads to a complete system with corresponding methods and processes.

From 1990 to 2020, some leading manufacturing companies in Europe, North America and Japan have established a cost engineering department, some of which are called Value Analysis & Value Engineering (VAVE), in Japan called Original Price Planning. Daimler has a cost engineering department of about 700 cost engineers, and Toyota and Nissan have similar departments of hundreds. Suppliers such as Lear, Delphi, Kostal, Hella, etc. also have cost engineers, but there is not yet a fully established cost engineering system (see Fig. 2.1).

Cost engineering is effective in product definition, optimizing product design, selecting the best materials and manufacturing processes, controlling design changes, optimizing logistics costs, etc., in order to reduce costs, improve the company's profitability and increase the company's competitiveness in the market, while ensuring product function and quality.

In China, many manufacturing companies have no concept of cost engineering at all and are not sure what the cost engineering department does. What is cost engineering? There were even counter-questions from the management to the keyword "cost engineering": "Do you mean cost? Or do you mean engineering (development or technical solution)?".

There are few joint ventures or foreign companies that have very small cost engineering teams, but these companies basically have no R&D departments in China or the existing domestic R&D departments do very little R&D work and have little authority to make product decisions, so such cost engineering teams are not very useful in practice.

X. Liu, *Cost Engineering Systems for Manufacturing Enterprises*,
https://doi.org/10.1007/978-981-96-5234-1_2

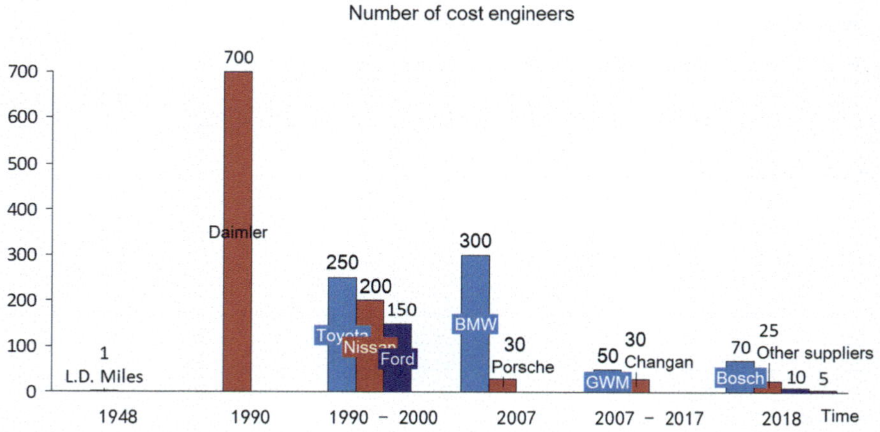

Fig. 2.1 History of cost engineering in the automotive industry and size of the cost engineering department

At present, cost engineering in China is still in the early stage, and Chinese manufacturing enterprises still need to invest a lot of work in cost engineering to effectively reduce costs and improve profitability, and they need to establish a complete cost engineering system within the enterprise organization and introduce systematic cost engineering methods and processes.

In recent years, domestic manufacturing enterprises have developed rapidly, and the awareness of forming a professional cost engineering department has awakened. However, the majority of component manufacturing enterprises still do not have a deep understanding of the importance of cost engineering in the product process and its contribution to the enterprise, believing that "cost engineering is dispensable to the enterprise, you do a little when you think of it".

To be a qualified cost engineer or cost engineering manager, you need a deep understanding of customer needs, experience in product development and manufacturing processes, strong interdepartmental and interdisciplinary coordination ability to move things forward.

Cost engineering is not just about costing or bid evaluation, nor is it just about lean manufacturing, but about working with R&D, finance, purchasing, production and quality to form a simultaneous engineering team that works on product function/performance, cost, production and quality to achieve the best possible result.

A cost engineering system with appropriate methods and processes is an essential part of the business framework for manufacturing companies and is indispensable. The cost engineering system is capable of reducing product costs by 5–15%.

There are three core tasks for production companies in the future: product innovation, intelligent manufacturing and cost reduction. Cost engineering goes beyond cost reduction and provides a solid and reliable basis for optimizing product design, manufacturing processes and improving customer satisfaction.

2.1 Cost Engineering in the Practice of European Manufacturing Companies

Most manufacturing companies in Europe initially still manage their costs as data. In the finance department, cost data from various cost centers was needed to calculate the net margin for a specific product or the income and expenses of the entire company; the finance department had a cost accounting team.

In the purchasing department, when obtaining a quotation from a supplier, the appropriateness of the quotation must be analyzed and a cost analysis of the quotation breakdown must be carried out; there is a cost analysis group in the purchasing department.

There are three elements to a product development project: quality, cost and time. The need to manage and report on the cost status of the project, which the project manager alone cannot do, a cost manager is brought in for the project.

In the sales department, a quotation must be submitted to the customer, and personnel capable of calculating the cost of the product are needed first; therefore, product cost estimators are added to the sales department.

Costs are primarily associated with summarizing, analyzing, managing and controlling data. On the contrary, costs are most associated with the technology related to product design and manufacturing.

It was not until the 1990s that costs were slowly linked to technology in European manufacturing companies. In 1990, Daimler set up a separate "Preliminary and Cost Planning" department, which was later renamed "Product Cost Engineering" and was responsible for cars, vans and trucks. It is the third largest department after Production and Research and Development and employs around 700 cost engineers.

Daimler's Product Cost Engineering aims to reduce product costs through technical optimization in the phases of product definition, product development and production while maintaining product quality and customer functionality, and currently has a fully integrated cost engineering system with the methods, processes and cost database embedded in the product process (see Fig. 2.2).

The main tasks of the Product Cost Engineering department at Daimler are.

(1) Calculation of the reference costs of components (REC0: zero-level reference cost) on the basis of technical designs and relevant data and forwarding them to the purchasing department as target costs for purchasing; support in commercial negotiations for purchasing.
(2) Module teams, composed of the three core members R&D, cost and quality, are responsible for technical designs, cost and quality for all vehicle series; they design modules (subsystems) that meet cost and quality targets by working synchronously across disciplines and departments.
(3) Perform cost change assessments and calculations (including component and tooling costs) for each engineering change.
(4) The Product Cost Engineering department is also responsible for analyzing competitive products, tool planning and building a comprehensive cost database.

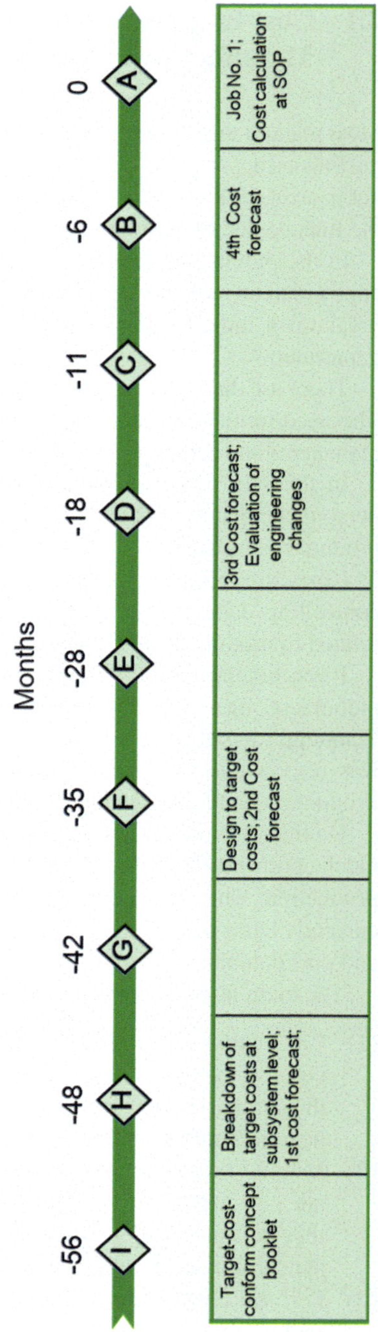

Fig. 2.2 Cost engineering embedded in the product process with corresponding milestones

(5) At the start of series production, the Product Cost Engineering department calculates the REC1 (first level reference cost).

BMW has also set up a cost engineering department since 2006 and currently employs around 300 cost engineers.

Parts manufacturers entered cost engineering relatively late, and most of them are still at the cost management level, where they implement some necessary measures to reduce costs and increase efficiency. Some large supplier companies do not even have a dedicated cost management department (they have cost estimators in their sales departments or cost managers in their finance departments).

At the beginning of 2018, Bosch Germany decided to set up cost engineering departments for each division.

2.2 Original Price Planning in Practice in Japanese Manufacturing Companies

The word "original price" is Japanese for "cost". Toyota Motor Corporation, for example, has established a department for original price planning.

The Product Definition Department, which is part of the Sales Department, defines the functions and features required for new products, focusing on the market and the customer, and provides this data to the Product Determination Department, which is part of R&D (see Fig. 2.3).

Product definition creates technical plans for new products based on input from the product definition department, while initial price planning sets the initial price target.

A chief engineer is appointed who is responsible for the entire new product, including customer features, initial price, quality and schedule. During the development phase of a new product, several meetings are held with the company management to review the original price situation.

Different original price achievement rates are set at different times, e.g. the first original price review must achieve an original price of 95% of the original price target, the second review must achieve an original price of 97.5%, and the SOP must achieve an original price of 100% of the original price target.

Toyota has its own original price planning department with hundreds of employees. Toyota places great emphasis on reducing the original price of mass-produced products and eliminating waste, and these efforts are integrated into Toyota Production System (TPS).

Nissan has a similar department for original price planning.

Parts suppliers such as Denso also have original price planning staff who are responsible for estimating original prices in the early stages of parts development, supporting sales in preparing quotations for customers, and carrying out projects to improve the original prices of parts. This is the main task of most component manufacturers in original price planning.

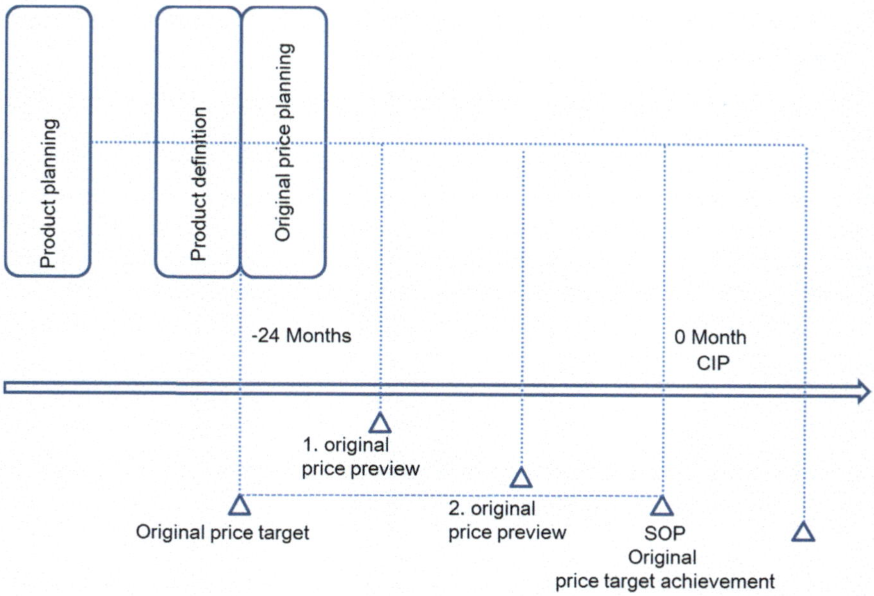

Fig. 2.3 Toyota's product planning, product determination and original price planning

2.3 The Practice of Cost Estimators in American Manufacturing Companies

In this example of the Ford Motor Company, the cost estimator is subordinate to the finance department, and the main tasks of the cost estimator are.

- Selection of "star parts", i.e. parts with a relatively high cost reduction potential and a high probability of realization in technical and commercial terms
- Calculation or estimation of the costs of these star parts
- Cooperation with Purchasing to negotiate technical and commercial issues with suppliers and achieve cost reductions.

At General Motors they are called Cost Engineers and their main tasks are

- Estimating the cost of components based on the technical design of the product, which serves as the basis for setting target prices for purchasing
- Setting component cost targets in the early stages of product development
- Evaluation of cost changes for ECRs (engineering changes)
- Supporting R&D and product project teams in implementing cost reduction programs during the development process.

Parts manufacturers have a similar division of labor to Japanese parts manufacturers, where the main role of the cost engineer is to make cost estimates at an early stage of parts development and support the sales department in pricing and optimizing parts costs.

2.4 Cost Management in the Practice of Chinese Manufacturing Companies

Before introducing the current practice of cost management in Chinese manufacturing companies, a personal experience of the author is described.

In 2018, the author had a conversation with the management of a large Chinese, on stock exchange listed component manufacturer, and the topic of the conversation was "reducing the cost of components".

"Why should we reduce the cost of parts, the results are passed on one-to-one to OEMs as it is, without us sharing any of it? We also have to invest in the design and validation of the modified parts", there is no motivation for reducing costs. That was a very impressive statement for the author.

The word "cost" is still seen as "managed" in most manufacturing companies. Many manufacturing companies do not have a cost engineering department, and costs are usually recorded in the finance, project or procurement departments.

A German colleague who was responsible for cost engineering in a Chinese-German joint venture, once told the author about an incredible experience over the phone. During a negotiation with a supplier, he introduced himself personally: "I am responsible for cost engineering", whereupon the supplier asked in astonishment: "Are you responsible for cost or engineering (development, technical solution)?" He replied: "Cost engineering, cost and engineering are written together". It took a while for the supplier to understand what his job was: "Cost engineering is the use of technical means to optimize costs".

It is time for Chinese manufacturing companies to wake up. Cost and technical solution (product design and manufacturing) are one to one, hence the name cost engineering, and product cost is not managed but designed and manufactured! The optimization of cost is the optimization of product design and manufacturing process.

In recent years, some Chinese manufacturing companies have begun to recognize the important role of cost engineering for their businesses.

Great Wall Motor, for example, commissioned the author to set up a brand-new department for original price planning and lightweighting in 2015 as part of the reform of its R&D center. Great Wall Motor's original price planning department is under the development center and has the following main tasks (Table 2.1).

Great Wall Motor's original price planning department, which led the Group's "Total Elimination of Waste" project in 2016, achieved cost reductions of CNY 400 million in 2016 and CNY 680 million in 2017.

Chang'an Auto has an institute for original price planning, BYD Auto has a cost engineering team, and Geely Auto started to install an independent cost engineering department in 2020.

Table 2.1 Tasks of original price planning

Work content	Remark
Business case calculation (NPV) and sensitivity analysis for new products	Inputs: forecast sales volume, forecast sales price, target net margin, etc.
Determination of target costs and target options	List of options, estimate of costs for new options
Allocation of the target costs of the product to the individual subsystems	Initial-BoM
Design to cost	Before freezing the data
Cost evaluation of engineering changes	Technical solutions for engineering changes
In series production	VAVE

Most component manufacturing companies still do not have a fully established cost engineering system, but only selectively use cost engineering, such as cost estimates needed for quotations from the sales department or to support R&D departments in component cost reduction projects.

The author expects that more and more Chinese manufacturing companies will quickly establish a fully integrated cost engineering system that meets the standard of the cost engineering system, improves corporate profits and competitiveness in the international market, and safeguards or increases the number of high value-added jobs.

Chapter 3
Product Function and Product Value

3.1 The Origins of Value Management

In the late 1940s, the devastation caused by the war, the lack of resources and manpower caused professionals and managers in manufacturing and service companies to rethink their product design or the composition of their services, opening up a new side to the question of how to provide adequate customer value.

In the fall of 1948, Lawrence D. Miles, a General Electric engineer (more specifically, a purchasing manager), first proposed that the function of products or services must be aligned with their value.

A product or service is generally considered valuable if its performance and cost are right. Conversely, a product that does not have an appropriate performance or price is not considered valuable. A scientific approach to analyzing and optimizing the cost of a product or service in terms of design (engineering), materials (labor) and manufacturing (process) has led to the concept of Value Analysis andValue Engineering (VAVE) (see also the book by Lawrence D. Miles "Techniques of Value Analysis and Engineering" [2]).

It has become the benchmark for good manufacturing and service companies to achieve the same functionality and performance at the lowest possible cost while maintaining quality.

Lawrence D. Miles, suggested that "value is always increased by reducing costs while maintaining performance".

3.2 The Relationship Between Product Function/ Performance and Product Value

We start with an analysis of the functions of a watch and the value of a watch, here a quartz watch is chosen as an example for the analysis.

X. Liu, *Cost Engineering Systems for Manufacturing Enterprises*,
https://doi.org/10.1007/978-981-96-5234-1_3

The functions of a product can be divided into main function, secondary function and image function, and each "function" has a corresponding "function carrier" that realizes this function.

For example, the main function of this quartz watch is the generation of time intervals, the display of time intervals and the attachment of parts (see Fig. 3.1), where the functional carriers for the generation of time intervals are the quartz and the battery, while the function of displaying time intervals requires the functional carriers stepper motor, dial and hands, and the function of fixing parts requires the case, the glass cover and the bracelet as functional carriers; the secondary function of this quartz watch is the protection function, and its functional carriers are the case and the glass cover; the functional carriers of the image function are embodied by the dial, the case, the glass cover and the bracelet.

The main function of this quartz watch was realized with 75% of the total costs, the secondary function with 10% of the total costs, and the image function with 15% of the total costs.

Depending on the customer group and the price, the engineer must select the functional carriers that are appropriate to the price the customer is willing to pay when implementing the main, secondary and image functions, which is the "target cost process" described in more detail in the following sections.

The analysis method described above is the function-cost-matrix method in cost engineering.

A car can drive on the road, that is its driving function as a car product; it can drive on the highway at a constant speed of 120 km/h, that is its driving performance;

It can accelerate when the driver presses the accelerator pedal, that is its acceleration function; if it can accelerate from 0 to 100 km/h in 4.5 s, that is its acceleration performance;

If it can brake to a standstill when the driver presses the brake pedal, this is its braking function; if it can brake from 100 km/h to a standstill within 35 m, this is its braking performance;

If the driver can steer the car with the steering wheel, this is his steering function; how fast he can turn in, this is his steering performance.

Based on the above analysis of the function-cost matrix approach for quartz clocks, in the case of passenger cars, the classification of "performance" is added to the function category, which also applies to the function-cost matrix approach (see Fig. 3.2).

In order to achieve the function/performance of a product, corresponding function/performance carriers with corresponding costs are required. The value of the product is inversely proportional to the cost of achieving the function/performance; the lower the cost of achieving the function/performance, the higher the value:

$$Value = \frac{function\ or\ performance}{Costs\ of\ the\ function\ or\ performance\ carrier} \tag{3.1}$$

From the customer's point of view, the value has two parameters.

Fig. 3.1 Table of functions and function carriers of a quartz watch

function-cost-matrix method (example: a quartz watch)						
function category	main function			secondary function	image function	total costs (CNY)
function carrier / function	generate time intervals (CNY)	display time intervals (CNY)	fastening parts (CNY)	protection grant (CNY)	image lending (CNY)	
clockwork (CNY)	42.12	4.68				
dial (CNY)		7.61			4.10	
pointer (CNY)		7.80				
housing (CNY)			5.85	7.80	5.85	
glass lid (CNY)			1.95	3.90	1.95	
bracelet (CNY)			5.85		5.86	
battery (CNY)	11.70					
total (CNY)	53.82	20.09	13.65	11.70	17.76	117.02
cost share in percent (%)	46%	17%	12%	10%	15%	100%

Fig. 3.2 Function/performance and function/performance carrier of a passenger car

Function-cost-matrix method (example a passenger car)

Function category	Main function					Main performance					Secondary function /performance	Image function	
Function carrier / **Function**	Drive (CNY)	See (CNY)	Acceleration slow down (CNY)	Brake (CNY)	Steering (CNY)	Driving at constant speed (CNY)	Acceleration performance (CNY)	Braking power (CNY)	Steering power (CNY)	Safety (CNY)	Comfort (CNY)	Image lending (CNY)	**Costs**
Engine (CNY)	5000.00		2000.00			1000.00	3000.00						
Transmission (CNY)	3000.00		1000.00			3000.00	1000.00						
Chassis (CNY)	2000.00		500.00			1000.00	500.00						
Body shell (CNY)	4000.00		1000.00			1000.00		1000.00		4000.00			
Seats (CNY)	1000.00		500.00			1000.00				500.00			
Braking system (CNY)				1000.00				1000.00					
Steering system (CNY)					2000.00				500.00				
Lighting system (CNY)		5000.00											
Heating and air conditioning system (CNY)											3000.00		
NVH measures (CNY)											500.00		
Decoration parts (CNY)												5000.00	
total (CNY)	15000.00	5000.00	5000.00	1000.00	2000.00	7000.00	4500.00	2000.00	500.00	4500.00	3500.00	5000.00	55000.00
Cost share in percent (%)	27%	9%	9%	2%	4%	13%	8%	4%	1%	8%	6%	9%	100%

- The content of the value
- The meaning of value.

This means that a reduction in cost for the same function/performance means an increase in value, often referred to as the price/performance ratio of the product.

The function of a product, which is the primary need of the customer, has only two states: present or absent (0 or 1).

And the performance of a product is the customer's secondary need. The performance of the product lies in a range between 0 and 100%. There are certain limits to what the customer is looking for in terms of the product's performance, e.g. price. If the cost of achieving a certain performance value is too high, resulting in a market price that is too high, the customer may choose to reduce or abandon this performance indicator, as this performance is irrelevant to the customer.

The main functions of a passenger cardriving, accelerating, braking and steering—are the customer's primary requirements for a passenger car and are realized by the corresponding function carriers (costs).

The customer's needs in terms of secondary performance such as acceleration, braking, steering performance, etc. are secondary. If the car is designed at a higher cost in order to achieve a certain performance target (e.g. accelerating from 0 to 100 km/h in 4.5 s) and the target customer does not consider this performance to be important for the customer, the costs are no longer justified.

The more important the functions or services are to the customer, the more resources and costs should be concentrated on these features, and conversely, the corresponding resources and costs should be reduced or eliminated.

3.3 The Essence of Value Management

Value management is a goal-oriented approach that uses suitable methods to implement and optimize products, work processes and services holistically and effectively, taking into account human behavior, the environment and corporate management.

Value management is therefore a holistic concept that encompasses the following key disciplines.

- Value analysis (value analysis and value engineering)
- Design to cost (Design to cost)
- Reference cost calculation
- Benchmarking
- Simultaneous engineering
- Life cycle management
- Modular strategy
- Optimization of the production system.

The core elements of value management span the entire product life cycle, from design to target cost, simultaneous engineering and modular strategy in the early stages of development to production system optimization and product life cycle management:

(1) A function-based observation approach
(2) The technical solution of the product is decomposed and abstracted on the basis of functions, listing the costs corresponding to each function, which are the core components of the product requirements
(3) Organized, systematic creativity: Through a systematic creative approach, the team will develop unconventional, new technical solutions
(4) Interdisciplinary, cross-departmental teamwork: The team consists of interdisciplinary and cross-departmental employees, not only from research and development, but also from cost engineering, quality, sales, customer service, production and purchasing. Complex problems are systematically broken down and abstracted, and team members form complementary relationships based on their knowledge and experience to achieve a comprehensive solution
(5) A systematic way of working: The work plan for value management, including each step of the work plan, should be implemented in an orderly and deciphered manner; the various methods and tools of the work plan should be applied precisely; and the entire work plan should be transparent to the teams involved and the decision-making level
(6) A networked mindset: Ensure interdisciplinary and cross-departmental thinking and action to achieve customer-oriented company goals
(7) Quantitative targets: Cost and quality targets must be quantified and the functional and quality characteristics of customer requirements must be achieved with minimal use of resources and costs; in the triangle of quality, cost and time, the purpose of value management is to optimize value (see Fig. 3.3)

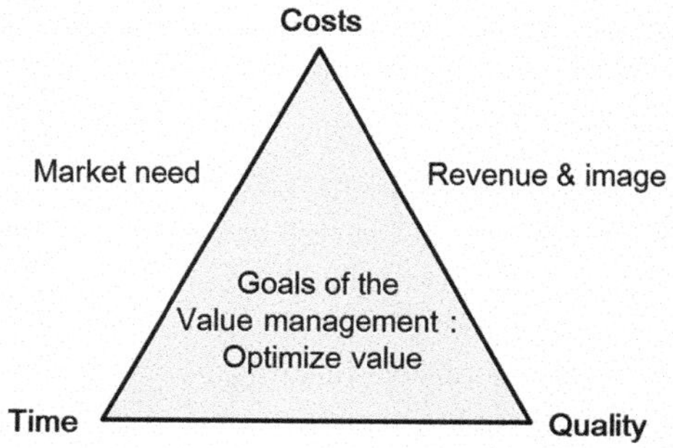

Fig. 3.3 Purpose of value management

Chapter 4
The Product Process

The product process is the entire process of product definition, development, production, sales and after-sales service, and up to the end of the product life cycle.

The product process also determines the cost of the product itself, especially at the beginning of the product process (product definition and product design), when 85% of the product cost is already determined, with a margin for cost optimization of about 15% for later stages of the product process.

The author will describe the product process in detail to enable the reader to understand the main sources and causes of cost generation and then apply the cost engineering methods described in the next chapters to systematically optimize product costs.

4.1 Definition of the Product Process

The product process consists of the following phases (see Fig. 4.1).

- Product definition phase
- Product development phase
- Product manufacturing phase
- Product sales and customer service phases.

1. The product definition phase

Product definition is the first phase of the product process and includes the following tasks:

(1) Target customer analysis

When a new product is launched on the market, what functions/services do customers need from the product? How important are these functions/services for the customer?

X. Liu, *Cost Engineering Systems for Manufacturing Enterprises*,
https://doi.org/10.1007/978-981-96-5234-1_4

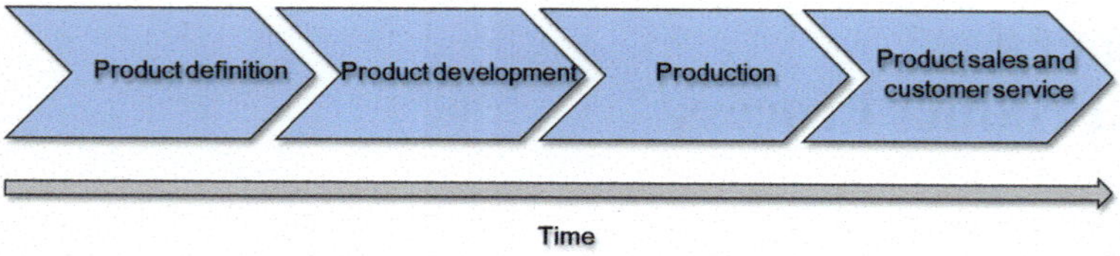

Fig. 4.1 Phases of the product process

How satisfied are customers with the existing product? Figure 4.2 illustrates the typical functions/services for a car and how important they are in the eyes of the customer perspective frequency of customer responses, where 100% is the highest value.

Mercedes-Benz vehicles are mostly perceived by customers as "safe", "comfortable" and "luxurious", BMW vehicles as "sporty", "good handling" and Toyota vehicles as "reliable" and "economical".

These labels are genuine customer ratings and are even decisive for the choice of a particular vehicle model. These are also the labels or brand values that manufacturers must uphold through research and development, sales and after-sales service, and are also the focus of manufacturer investment.

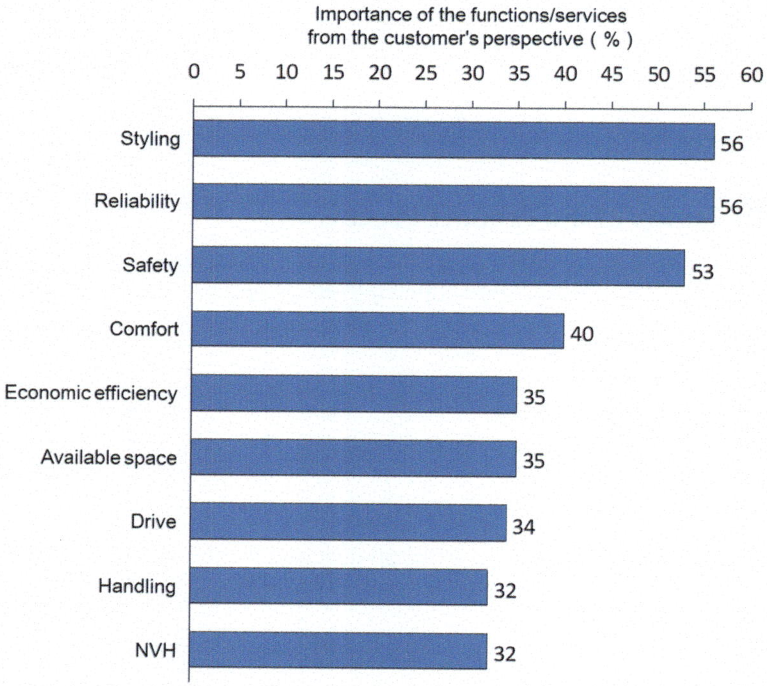

Fig. 4.2 Importance of the functions/services of a car from the customer's perspective

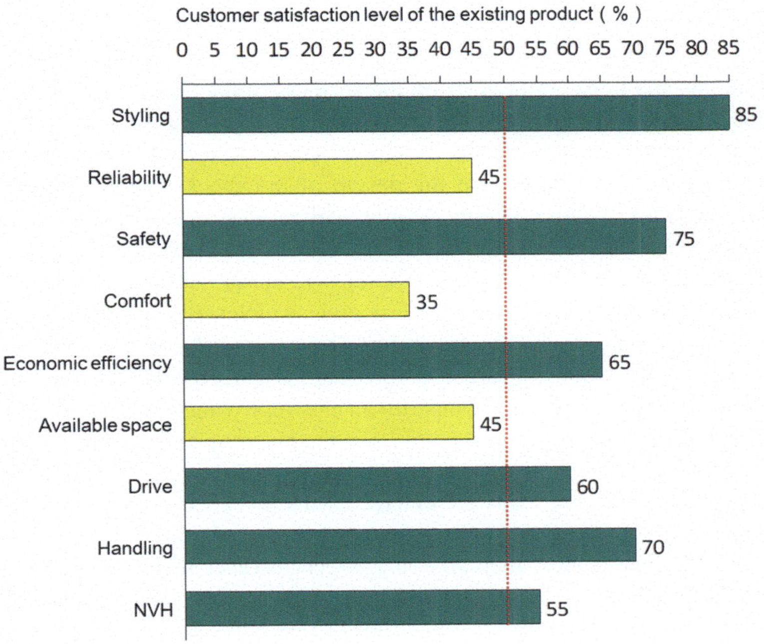

Fig. 4.3 Results of the customer satisfaction survey for an existing product

At the same time, customer satisfaction surveys with existing products are used to identify shortcomings and improve them.

Based on customer satisfaction surveys with existing products, a score of over 50% indicates that customers are satisfied with the features/performance of the product, while a score of less than 50% indicates that they are less satisfied.

The results of the customer satisfaction survey for an existing product in Fig. 4.3 exemplify that customers are satisfied with styling, safety and economy, while they are dissatisfied with reliability, comfort and space.

Improvement targets need to be set for each of these dissatisfied customer functions/services.

Reliability: The existing product has 350 individual failures or customer complaints per 1000 units, the improvement target is 250 individual failures or customer complaints, which corresponds to an increase in reliability of 28.6%. The technical causes of each individual failure or customer complaint and the subsystems affected are analyzed and the technical solutions for the subsystems concerned are developed for the new product in order to reduce or avoid failures or customer complaints, which is in line with the principle of design for quality.

Comfort, the general subjective feeling of a customer when using a product, must be measured analytically. It includes the overall vehicle styling, the seats, the powertrain, the chassis and other systems.

The size of the vehicle, the ergonomic design of the HMI (human–machine interface), the upholstery and adjustability of the seats, the ease of operating the gearshift, the vibrations and noise in the interior, the stiffness of the shock absorbers—all of these influence the customer's subjective perception of comfort.

The most important subsystems that affect comfort, such as the seat trim, the transmission control software and the design of the shock absorber springs, are examined in order to determine the development targets for the new generation of subsystems.

Space, divided into cockpit space and luggage space, is the space used by the customer, which affects the customer's functions, which serves the purpose of traveling and transporting goods, and its comfort. Determining the overall dimensions of the vehicle (length, width and height), improving the space available in the front and rear seats, enlarging the luggage compartment and improving the ease of loading and unloading luggage become important tasks in the planning and dimensional design of the new product.

All these improvement ideas are entered in the so-called product specification document.

Parts manufacturers must also carry out a target customer analysis, with the difference that the product is not delivered to the end customer, but to the next stage of the production company. Does the customer need to be satisfied with the reliability of the product? Are the requirements for reject rates (ppm values) appropriate? Does the customer consider the price to be too high, so that the company itself has to optimize costs? Or are there certain issues in the customer's product specification that are outside the normal scope of use? And so on.

By analyzing the target customer, the manufacturer can determine the future product definition and development direction.

(2) Product and technology roadmap

Which technological innovations are ready for the market from the first day of product definition to the time of market launch? Which of these economically viable technologies are suitable for use in the defined new product?

As can be seen from the example in Fig. 4.4, the 6-seater van concentrates on a customer-oriented technology path that focuses on the interior, the human–machine interface and the options.

The importance for the customer is the guideline for the technology path, such as "removable seats", "interior lighting", "inductive electric tailgate", "pedestrian protection", etc.

The "First Customer" department organized a prototype evaluation for the "removable seat", which was mounted on a fixed aluminum rail in the font area and tried out by engineers from various departments, male and female, sometimes stronger and sometimes less stronger.

From opening the locking switch, folding the backrest and the grip points to removing the seat, the "First Customer" department listened to various suggestions for improvement from colleagues and repeatedly optimized the weight of the seat,

Technology	Description	target vehicle	development time window		customer importance	remark
Interior and HMI						
Integrated storage solution for the dashboard	Straight dashboard reinforcement carrier; modular design in the center area for easy installation of navigation systems, tachograph, etc.	car line1			50	
CO2 air conditioning	CO2 as working medium	car line1			50	
New air conditioning concept: improved driver comfort (first and second stage)	Multifunctional air conditioning unit without draught (with multi-point, decentralized air outlet); design based on flow simulation	car line1			50	
New solution for air conditioning:	Compact defrosting	car line1			50	
Pre- or stationary air conditioning	Static air conditioning solution for greater comfort when entering the cabin	car line1			25	
Separate HMI and climate control	Standardization and user-friendliness of the instrument cluster	car line1			23	
Removable seat	Improved customer friendliness and comfort	car line1			75	
Seat with integrated small table top	Simplified use through font passages	car line1			50	
Variable rear seats	Designed in variable shapes for sitting and reclining positions	car line1			25	
Equipment						
Ambient lighting in the interior	Improved visibility, comfort and safety for passengers entering and exiting the cabin at night	car line1			75	
Interior lighting	Integration of different LED colors, lighting methods	car line1			25	
Passenger side entry handle	Easier access for passengers	car line1			50	
Roll-over airbag	Protection of passengers in the event of a rollover	car line1			25	
Family and child-friendly interior	Consisting of several concepts	car line1			50	
Electronic automatic sunshade	Infinitely variable sun protection	car line1			50	
Inductive electric tailgate	Convenient use of the rear door	car line1			75	
New electric sliding side door solution	Stepless stopping	car line1			75	
LED rear light		car line1			25	
Laser welding of body parts	Improved body rigidity and reduced vehicle weight	car line1			25	
Pedestrian protection	New design of the front bumper and hood	car line1			75	legal requirement
Lightweight body shell	Mixed construction of steel, aluminum and plastic	car line1			25	
Cd value (aerodynamically) optimized exterior	Optimized lower body cladding, front and rear of the vehicle	car line2			75	Reduction of fuel consumption
Electrochromatic glass	Protects and increases comfort in the interior	car line1, car line2			50	

Fig. 4.4 Technology path of a 6-seater van

the seat rails, the locking switch, safety and the target cost requirements in order to finally develop a solution for series production.

(3) Benchmark analysis of the competition

By comparing the relevant functions/performance of competitor products with the customer's requirements, a clear market positioning of the company's own product can be derived. If it still lags behind the competitor in some customer requirements, as in the case of Fig. 4.5 Competitive comparison functions/performance, and in terms of reliability, comfort and space, select the appropriate approaches in the technology path after the comparison and write these approaches in the product definition document.

(4) Option list

The option list is prepared by the sales, R&D, cost engineering and finance departments after conducting the target customer analysis, technology path and competitive comparison to meet the function/performance requirements of the target customer and the competitiveness of the product in the market, and the equipment list is also a part of the product specification document (Fig. 4.6).

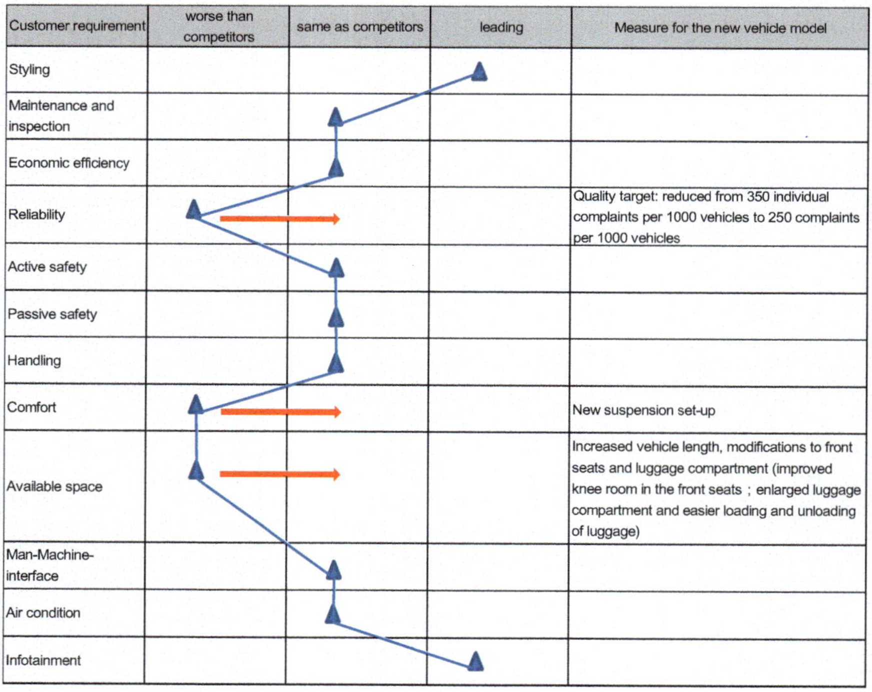

Fig. 4.5 Competitor comparison functions/performance

Option name		Entry Line	Medium Line	High Line	Top Line
Powertrain		1.6 L Turbo-charger + 7DCT			
1. Chassis					
1	MacPherson independent front suspension	●	●	●	●
2	Rear double wishbone independent suspension	●	●	●	●
3	Electric power steering	●	●	●	●
…	…	●	●	●	●
2. Exterior					
1	Door handles in the body color	●	●	●	●
2	Electric tailgate	×	○	○	●
3	Integrated spoiler antenna	●	●	●	●
…	…	●	●	●	●
3. Interior					
1	Steering wheel with genuine leather	●	●	●	●
2	Electrically heated steering wheel	×	×	●	●
3	Four-way manually adjustable steering wheel	●	●	●	●
…	…	×	●	●	●
4. Seats					
1	Business Class Seats	●	●	●	●
2	Faux leather and imitation suede seats	×	●	●	×
3	Seats with genuine leather	×	×	×	●
…	…	×	×	●	●
5. Active and passive Safety					
1	Two front-Airbags	●	●	●	●
2	Front side airbags	●	●	●	●
3	Integrated window bag at the front and rear	●	●	●	●
…	…	○	○	○	○
6. Theft protection					
1	Intelligent key	●	●	●	●
2	Keyless-Entry	●	●	●	●
3	One-touch start function at the touch of a button	●	●	●	●
7. Multimedia					
1	9 zoll TFT color display	●	●	●	●
2	Telematics	○	○	○	○
3	12.3 inch digital instrument cluster	×	×	●	●
…	…	○	○	●	●
8. Lighting					
1	Electric height-adjustable headlamp	●	●	×	×
2	LED day running light	●	●	●	●
…	…	×	●	●	●
9. Air condition					
1	Automatic climate control	●	×	×	×
2	Two zone climate control	●	●	●	●
3	Dust filter	●	●	●	●
…	…	●	●	●	●
10. other options					
1	0.5kg fire extinguisher	○	○	○	○
2	Warning triangle	●	●	●	●
3	first aid box (for export)	○	○	○	○
…	…	○	○	○	○
Remark					
Symbol : × without ● standard ○ option;					

Fig. 4.6 Option list for a passenger car (extract of options)

During the product definition phase, the customer's function/performance requirements for the product are at the forefront, and the careful selection of the option list is an important part of sales, R&D and cost engineering.

For example, does the customer really need the 360-degree surround view equipment? In what situation will it be used? And how often? Can a panoramic sunroof, which is expensive and increases the weight and energy consumption of the entire vehicle, be replaced by a smaller normal sunroof?

Everything is analyzed and ordered, whether it is planning technology paths, analyzing competitors or creating option lists, always with the customer's functional/performance requirements for the product in mind, in order to achieve them at minimum cost so that the product can achieve the highest customer benefit.

Production companies therefore need a precise customer image, a clear technical development direction and a precise option list for the product, no more and no less. This is the quintessence of cost engineering.

2. The product development phase

The product development phase is the phase in which the product is developed from the design data and drawings through product validation and trial assembly until it is ready for series production. The first result of product development is the technical concept, which already accounts for around 85% of the product costs. Product validation provides results on the reliability and durability of the product. During trial assembly, it is checked whether the product can be manufactured in series according to the production plan and quality requirements.

Product development is carried out in strict accordance with the product development process. Figure 4.7 shows the entire development process for a new vehicle model.

The development process for a complete vehicle is divided into nine stages: I, H, G, F, E, D, C, B and A.

In phase I, often referred to as the strategy phase, the framework of the new vehicle model is defined. The result is the concept booklet, which describes a preliminary concept for the new vehicle model, and phase I lasts around six to nine months.

Phase H, a detailed description and feasibility analysis of the new model in terms of total vehicle size, total vehicle performance, option, target cost, target selling price, projected sales volume, product life cycle, the result is a mature vehicle specification sheet, phase H takes about 8 months.

Phases G, F, E, D, C and B, the entire vehicle styling, design, validation and preparation for series production, are also the phases of new vehicle model development, in which various cost engineering methods and procedures are applied to design the most cost-effective product, which takes about 42 months.

The reader may wonder whether it really takes that long. The time ultimately required depends on the depth of development of the overall vehicle (how many subsystems have to be newly developed) and also on the number of validation objects (functional validation, durability validation) and the depth of validation (two winter and summer tests required?).

Fig. 4.7 Entire development process for a new vehicle model

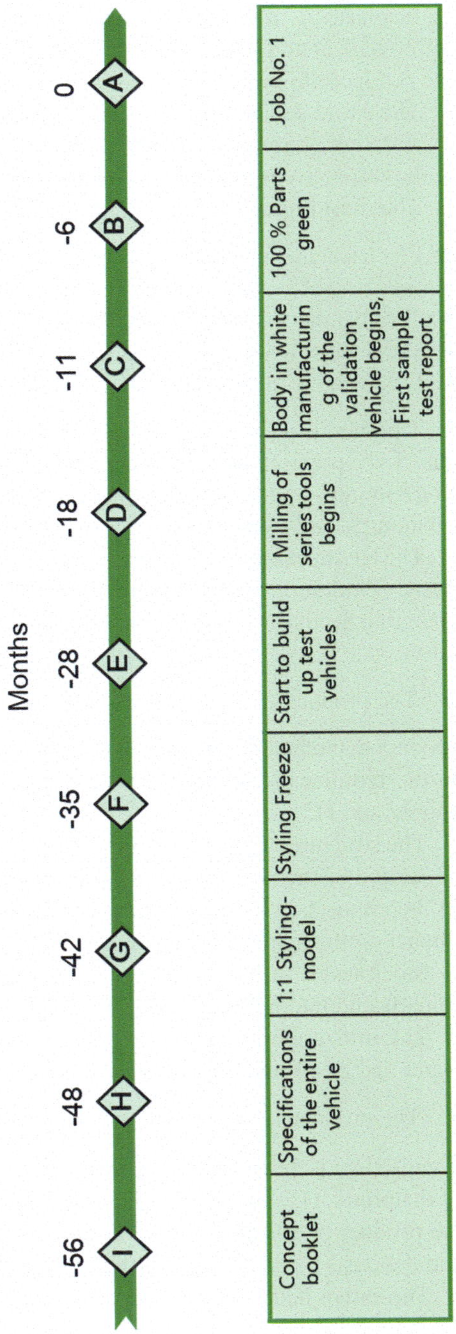

In phase A, the series production phase of a new vehicle model, continuous improvement and VAVE work is a "must" for production companies until the end of the product life cycle.

The entire development process is divided into a planning phase, a styling phase, a validation phase and a series production phase, each of which has a list of results at the respective milestones (see Fig. 4.8).

The most important points here are.

- Electrical and electronic architecture
- Freezing of styling and design data
- Toolmaking for series production
- 100% parts from series production tools
- First vehicle in series production (job no. 1).

The attentive reader will wonder whether it really takes 56 months to develop a new car? To be precise, the normal vehicle development cycle is 48 months, excluding the first 8 months of the strategic planning phase. Many automotive companies only need 36 months, and some even have a full vehicle development cycle of only 24 months.

The length of the development cycle naturally determines most of the development costs. Shortening the development cycle is also a way of adapting to the rapidly changing needs of customers and the market environment and reducing development costs.

3. The product manufacturing phase

Product manufacturing is carried out using a complete production system. In addition to the familiar production line, there are supporting systems for logistics, energy supply and IT.

The structure of the production line, the utilization of the production facilities, the allocation of the work steps to the individual stations, the arrangement and variety of the connecting elements and tools as well as the production cycle have a direct impact on the manufacturing costs.

Stock levels in the logistics area, the distribution of components, the packaging of parts and internal transportation are also part of the manufacturing costs.

The utilization efficiency of energy and other consumables such as electricity, water and gas is also a significant component of manufacturing costs.

4. The product sales and after-sales service phase

The product is delivered to the customer, and only then does the most critical phase of the product begin. Do the functions requested by the customer work properly? Is the product performing as the customer expects? The product requires a warranty and the costs incurred are part of the product cost.

The author had a personal experience when he was working in Germany: one day the author drove to work early in the morning to the Daimler Group Research Center, during the drive the brightness of the headlights diminished more and more and the car stopped and could no longer be started.

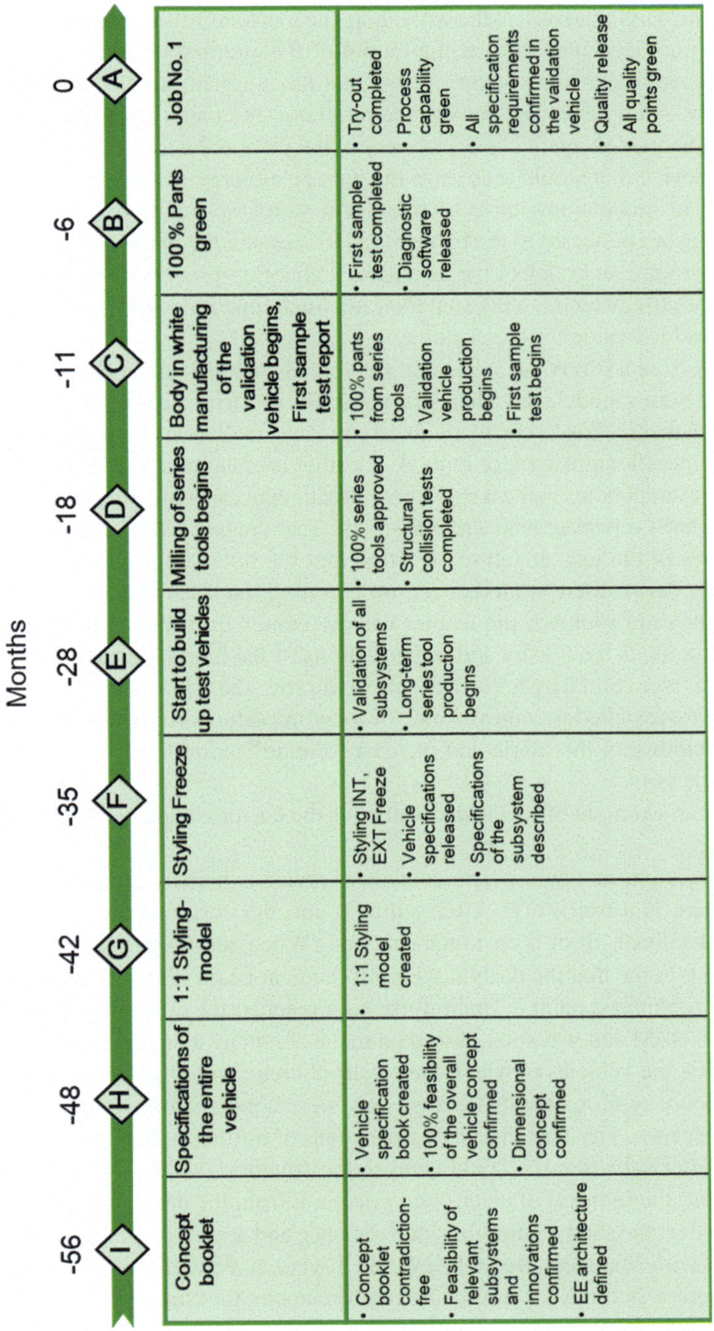

Fig. 4.8 Development process of a new vehicle model with corresponding results at the respective milestones

When the author arrived at the workshop, the mechanic discovered that the voltage regulator (an electronic chip) at the output of the alternator had failed, so that the car continued to consume battery power and the alternator could no longer charge the battery. After replacing the voltage regulator, the battery was charged and the car was able to start again. As a customer at the time, the author wished there was a warning device that would recognize the battery's energy reserves.

Some car models now have an intelligent start/stop function, which includes a current sensor connected to the battery lead to measure the battery's current, voltage and temperature and control the alternator's charging process, which also protects the battery's life, which is a big step forward. Such functions and the associated costs are a real added value for the customer.

A story from a BMW car salesman: a potential customer came into the 4S store to look at a 3 Series model, and when he touched the seat trim, he asked the sales rep, "Is this real leather?" "Yes", "Why are there so many small holes in the leather fabric?" To which the sales ambassador replied, "Leather is breathable and comfortable to sit on, and the small holes make it even more breathable, especially in summer, you will like it". That's a perfect answer. What is the real context? Why are there so many small holes in the leather fabric? It's true that the holes improve the breathability of the seat fabric; there's another reason here that the sales ambassador didn't tell (or perhaps didn't know): the leather usually comes from the cattle that are bred, which sometimes have scars and imperfections of the hide because they walked on the grass or were bitten by horseflies, and to improve the utilization of the cattle hide, the holes are punched to remove some of these imperfections, which both improved the breathability of the fabric and increased the utilization level of the leather, and reduced the cost.

This is an example of real added value for the customer and cost reduction at the same time.

In today's internet age, where customer service is heavily influenced by online media, there is a true story: After selling a car, the customer service department received feedback from a customer or two: "When servicing the car, the service employee told me that the dashboard reinforcement beam of my car was not coated with electrophoretic paint." This information reached the customer service department of an OEM and was soon passed on to the company management, who decided to take back the vehicles in which the electrophoretic paint had been removed from the dashboard reinforcements because they were concerned about the loss of sales this would cause. This cost the car company tens of millions of CNY more each year.

If the after-sales service could analyze the situation calmly, the situation might be different: the removal of electrophoretic paint from the dashboard reinforcement does not affect the safety, durability, performance and aesthetics of the whole vehicle; compared with Mercedes-Benz, BMW and Toyota, none of the vehicle models have electrophoretic paint on the dashboard reinforcement; the complaints and grievances of 1– 2 customers, not 100 or even 1000 customers, can only be said to be isolated cases; the after-sales service personnel can show the customer a patient technical explanation and comparison. Basically, the tens of millions of CNY this car company spends on this matter every year is no longer a real customer benefit.

The essence of cost engineering is to always keep customer requirements in mind, "turn over" every cent twice, and handle it very carefully to achieve real, high customer value at the lowest possible cost.

4.2 Responsibility Matrix for the Product Process (RASIC)

As shown in Table 4.1, the responsibility matrix for the product process defines the work and responsibilities of departments such as Sales/Customer Service, R&D, Cost Engineering, Quality, Purchasing, Production and Finance.

R: Responsible; A: Accountable; S: Supportive; C: Consulted; I: Informed.

4.2.1 Tasks and Responsibilities of R&D in the Product Process

The main tasks of R&D in the product process are.

- Technology path planning
- Analysis of competitor products
- Product definition
- Product design
- Product validation.

R&D is primarily responsible for technical product solutions and product reliability in line with customer requirements for product function and performance and supports the selection of suppliers.

4.2.2 Role and Responsibility of Quality in the Product Process

The main tasks of the quality department are.

- Defining the quality objectives of the product
- First sample test
- Release of the first try-out for series production.

The quality target is reduced from 350 individual customer complaints or problems per 1000 vehicles for a series product to 250 per 1000 vehicles (see Fig. 4.9), and the 250 individual customer complaints or problems are broken down individually to the corresponding subsystems.

Table 4.1 Responsibility matrix for the product process (RASIC)

Task	Sales and customer service	R&D	Cost engineering	Quality	Procurement	Production	Finance
Target customer analysis	R, A	C	C	S			I
Technology path planning		R, A	C	C	S		
Competitive analysis		R, A	S	C			
Forecasted sales volume and price	R, A		C				S
Quality objectives of the product		S	C	R, A			
Target costs of the product			R, A		C		S
Product definition	S	R; A	C	C	C	I	C
Product development		R, A	S	S	C	C	
Supplier definition		S	S	C	R, A	C	
Product validation		R, A			C	S	
First release of a try-out				R, A	S	C	
Engineering changes		R, A	S		C		C
Production start-up		C		S	C	R, A	
Cost calculation of the series product			S		C	C	R, A

If the product is a passenger car, there may be individual problems such as the vehicle not starting (battery flat), the door windows not closing, drive shaft resonance at a certain speed, excessive wind noise at high speed (in the wing mirror area), flap noise in the rear while driving, large noises when the air conditioning is switched on, etc.

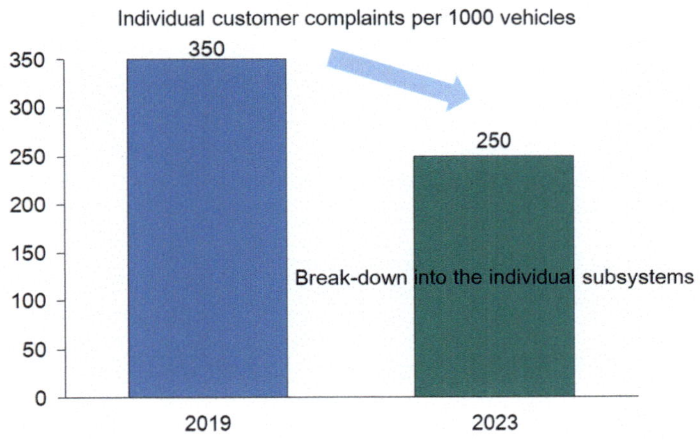

Fig. 4.9 Determination of the quality targets for a new vehicle model

The main role and responsibility of the quality engineer in the product process is to work with the R&D engineers and cost engineers to find technical design solutions or remedies to avoid these individual faults or problems, some of which are subsystem improvements and some of which are architectural improvements that need to be made at the entire vehicle level.

The initial sample testing, as well as the testing and approval of the first try-out prototypes, are important steps for the quality engineers to holistically test the functions/performance of the product and validate the production consistency of the product under series production conditions.

4.2.3 Tasks and Responsibilities of Cost Engineering in the Product Process

The main tasks of the Cost Engineering department in the product process are.

- Determine target costs for the product
- work with the R&D, Quality and other departments to implement design to cost for the product in a synchronized technical manner
- Evaluate and manage engineering changes
- Lead work to reduce costs and increase efficiency of series products.

The Cost Engineering department is responsible for determining the target cost of the product based on the company's sales forecast, sales price and profit margin target and allocating it to the subsystems; in the product development phase, they apply design-to-cost methods together with the R&D, Quality and Purchasing departments to ensure that the target cost of the product is achieved as planned.

Chapter 5
Cost Engineering Methods

There are dozens of methods for cost engineering, which are mainly divided into the categories of analysis, design and process/commercial issue.

Those belonging to the analysis category, such as value analysis, reference cost calculation and multi-dimensional ABC analysis.

Those belonging to the design category, such as the morphological box method, the cost-function-matrix method and the benchmark-function method.

Those belonging to the process category, such as the target cost process and lean production.

Those finally belonging to the commercial category, e.g. business case calculation (NPV), negotiation strategy & tactics.

In the following sections, the cost engineering methods mainly used in practice in manufacturing companies are described in detail and provided with corresponding practical examples.

5.1 Multi-dimensional ABC Analysis

ABC analysis assumes an order of components, usually based on the value of the cost of components arranged from the largest to the smallest, with limits set for the level of classification, e.g. more than 100 CNY for Class A components, more than 10 CNY and less than or equal to 100 CNY for Class B components, and less than or equal to 10 CNY for Class C components. The results of the ABC analysis of a passenger car are shown in Tables 5.1, 5.2 and 5.3.

With the ABC analysis, the top 100 with the highest cost values can be filtered out of thousands of components with a cost share of 75–85% of an entire vehicle, so that the main points can be recorded and a detailed cost optimization can be carried out.

X. Liu, *Cost Engineering Systems for Manufacturing Enterprises*,
https://doi.org/10.1007/978-981-96-5234-1_5

Table 5.1 Class A subsystems/components (≥100 CNY) 110 in total, accounting for 79% of the total cost of the vehicle

Component name	Quantity	Costs (CNY)	Classification
Engine	1	12413.75	A
Automatic transmission	1	8961.48	A
Rear axle	1	2226.85	A
Electric power steering	1	2222.40	A
Sunroof	1	2208.00	A
……	…	…	A
Oil cooler	1	100.00	A

Table 5.2 There are 423 category B subsystems/components (≥ 10 CNY and < 100 CNY), which account for 19% of the total cost of the vehicle

Component name	Quantity	Costs (CNY)	Classification
Ultrasonic radar control unit	1	99.20	B
Front guard	1	98.83	B
3rd rear brake light	1	98.59	B
AS Glove compartment	1	96.28	B
AS Side airbag	1	96.00	B
…	…	…	B
2.0L engine badge	1	10.13	B

Table 5.3 Total of 1016 category C subsystems/components (<10 CNY), which account for 2% of total vehicle costs

Component name	Quantity	Costs (CNY)	Classification
Bracket for the rear bumper	2	9.93	C
Rear floor cover	1	9.83	C
Lower protective device of the right-hand B-pillar	1	9.69	C
B-pillar, lower protective device	1	9.68	C
Heat shield	1	9.53	C
…	…	…	C
Bolt	2	0.02	C

The disadvantage of the ABC analysis is that there is only one dimension (the dimension cost of a component) and the following aspects are not taken into account:

- Market penetration
- R&D costs
- Tooling costs.

Market penetration is the number of suppliers available on the market for the part, and R&D costs and tooling costs are the R&D and tooling costs that need to be invested in the serial production of the part. If the costs of the component are assumed to be a C(i) function, the market penetration rate as an S(i) function, the R&D costs as a D(i) function and the tooling costs as a T(i) function, then the multi-dimensional ABC analysis function $M_A(i)$ is

$$M_A(i) = F_N \frac{C(i)S(i)}{D(i)T(i)} \tag{5.1}$$

where F_N is the normalization factor.

The horizontal coordinate i refers to the various components and the vertical coordinate is the value of $M_A(i)$ as a function of i.

The ranking of the results differs between the multi-dimensional ABC analysis and the simple, component-cost-based ABC analysis. Here, market penetration, R&D and tooling costs directly influence the value of the multi-dimensional ABC analysis function $M_A(i)$. . The multi-dimensional ABC analysis makes it possible to analytically identify components with high cost reduction potential (high cost values), low one-time costs (R&D, tooling costs) and easy market realization (high market penetration).

Figure 5.1 shows some of the results of a multi-dimensional ABC analysis for an SUV. The components at the top of the list are the ones that should be focused on for cost optimization and are also the most feasible.

The engine is in first place because of its high cost, and the assembly sunroof (ZB sunroof) is in second place because, although it is not very expensive, the one-time investment (research and development, tooling) is lower, so it is also at the top of the list.

Components, possibly Assemblies, such as tires, AS rear axle and AS wiring harness for the dashboard are ranked higher because they are subsystems with high cost reduction potential and low one-time investments. In contrast, sub-assemblies such as automatic transmissions have higher unit costs, but also higher one-time investment costs (validation, tooling), which puts them at the bottom of the list.

5.2 Reference Cost Calculation

The reference costs are the sum of the individual costs incurred in the manufacture of a component and are made up of the following components.

- material costs
- manufacturing costs
- Administrative costs (overhead).

The material costs include the costs for raw materials, the costs for purchased parts and the costs for logistics in the plant.

Fig. 5.1 Selected results of a multi-dimensional ABC analysis for an SUV

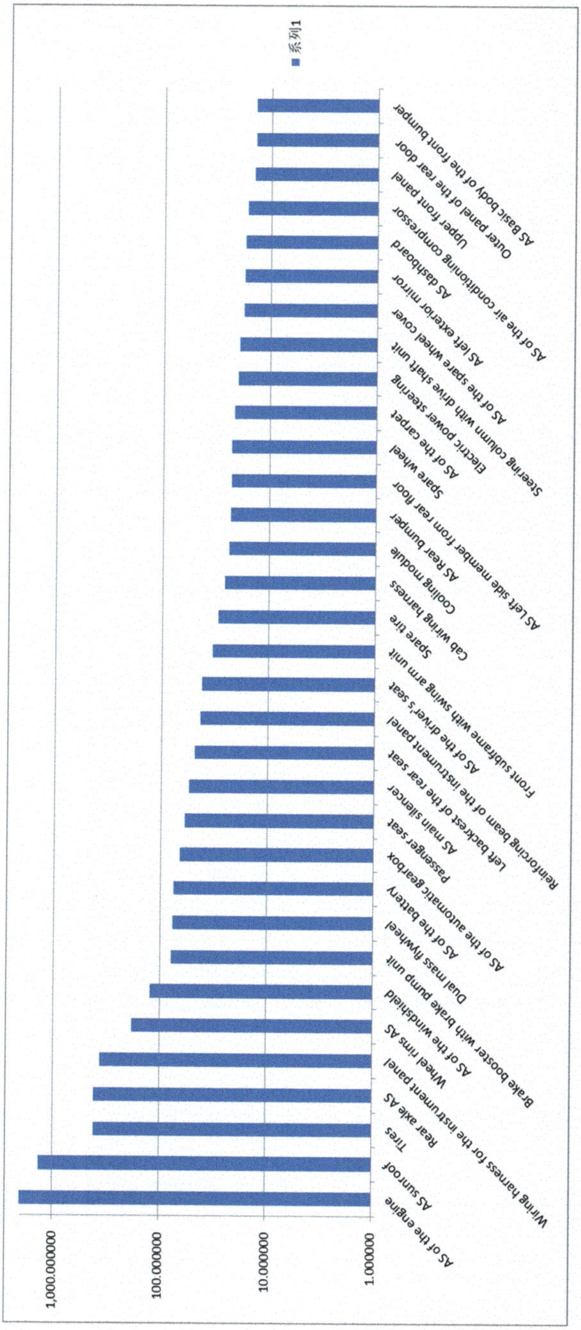

Manufacturing costs include labor costs, depreciation of plant equipment, energy costs and production overheads for each process step used to manufacture the component.

The administrative costs (overhead) include sales/administration costs, R&D cost amortization, profit, packaging and transportation, scrap rate, tooling cost amortization, etc.

The reference cost of a component can be calculated using a precise mathematical model that simulates the entire logistics, manufacturing and management process of the component at the benchmark level, also called green field calculation.

As shown in Fig. 5.2, the raw materials are transported via internal logistics to the production plant, where the raw materials are processed by the production facilities and the processed semi-finished products are assembled into finished products.

In the production facility, there are workers who are directly involved in production, quality inspectors and maintenance personnel who are indirectly involved in production, and the manager of the production facility.

The assembled finished products are then packaged and shipped to the customer via the logistics department. Costs are also incurred in the company's administration, sales and R&D departments, which must be allocated to the component.

The cost calculation is divided into two categories. The first category is the costs that are calculated according to the status of series production, regardless of whether it is in-house production or subcontracted production, and includes the construction of the production line, the process level and the management level of series production. The second category is the cost of a component, which is calculated on the assumption that the production of the component is at a benchmark level in the production and management. These calculated costs are called reference costs or green field calculation.

The following descriptions are reference costs. Taking a millimeter-wave radar cover as an example (see Fig. 5.3), the reference cost is calculated to be CNY 234.52 using the mathematical model described above, with the cost break-down as follows.

The calculation of the reference costs of the component leads to the following two results:

(1) Cost breakdown: this includes material costs, manufacturing costs and administrative costs (see Table 5.4). The result of the reference cost calculation can be used directly to evaluate the supplier's offer: Is the price within a reasonable range? In the example of millimeter wave radar coverage shown in Fig. 5.4, the offer from Supplier A is clearly outside the reasonable range, while the offers from Supplier B and Supplier C are within the reasonable range. This is a real case in which supplier A was no longer considered in the ordering process after the reference cost calculation and supplier B was ultimately selected.

Does the supplier also provide a cost breakdown? If so, is the cost breakdown also within the projected range? What about the production for each process step? The cost details determined using the mathematical reference cost calculation model, e.g. material cost details, manufacturing cost details, management cost details, can be used as arguments for business negotiations.

Fig. 5.2 Mathematical model for reference cost calculation

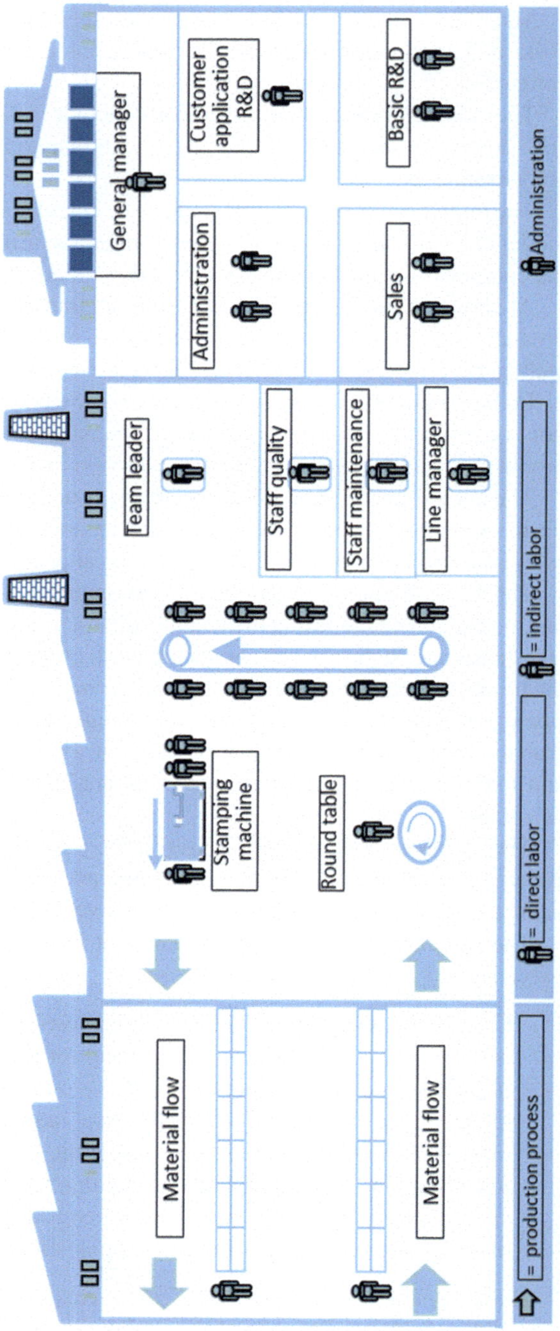

Fig. 5.3 Millimeter-wave radar coverage for a passenger car

Table 5.4 Reference cost breakdown for millimeter-wave radar coverage

Cost type	Value (CNY)
Reference costs	
Material costs	152.13
Material management costs	3.96
Production costs	44.03
Production overheads	6.48
Cost of sales, administration	14.46
Profit	5.68
Logistics, Transportation	2.57
Scrap	4.09
Total reference costs	234.52
Total material costs	152.13
Front cover (PC material)	5.68
Indium metal coating	141.04
Rear cover (AES material)	5.17
Black color	0.12
Protective color	0.12
Total manufacturing costs	44.03
Front cover (injection molding)	0.75
PVD coating (indium)	24.93
Rear cover (injection molding)	0.75
Painting	0.55
Assembly	0.50
Inline testing (EOL test)	16.55

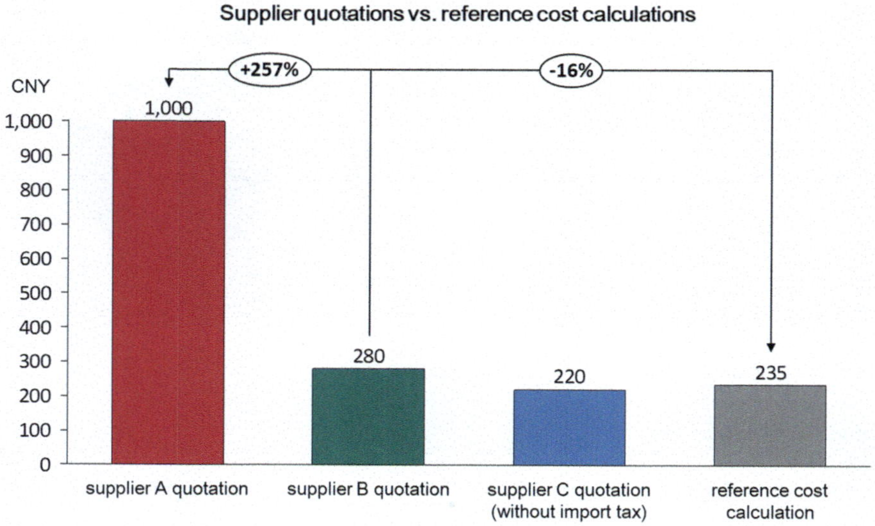

Fig. 5.4 Supplier offers versus reference cost calculation of millimeter wave radar coverage

(2) Cost drivers: Select the 3–5 largest cost drivers from the breakdown of material costs and manufacturing costs. Take the millimeter wave radar cover above as an example: Indium metal coating at 141.04 CNY, PVD coating process at 24.93 CNY and EOL test at 16.55 CNY are the 3 biggest cost drivers for the millimeter wave radar cover. These three cost drivers are also potential cost optimization points, e.g. can the thickness of the indium metal coating be reduced? How can the material utilization of the PVD coating process be improved? Can the inline inspection also be carried out offline (in batch mode) so that no time is taken up on the expensive production line? By optimizing the cost drivers, a cost reduction of 47.89 CNY was achieved, as shown in Table 5.5.

The advantage of the reference cost calculation is that it not only provides a breakdown of the costs of the component, but by analyzing the cost drivers it also finds the technical cost reduction points that are not obvious but lie deeper.

The reference cost calculation model is a general mathematical model that enables the calculation of different components or subsystems such as engines, transmissions, body-in-white, chassis, interior and exterior, wiring harnesses, electronic

Table 5.5 Optimized cost driver points for millimeter wave radar coverage

Cost driver	Cost reduction potential (CNY)
Indium-metal coating (35 μm reduced to 25 μm)	40.30
PVD coating (indium metal) (10% increase in coating efficiency)	2.49
EOL test (offline in batch mode)	5.10

Fig. 5.5 An electronic airbag control unit for a passenger car

control units, etc., taking into account different manufacturing processes (primary forming, forming, separation, joining, coating, change of material properties) and combinations of these manufacturing processes.

The following is an example of a reference cost calculation for a "high version" of an electronic airbag control unit (ACM) (see Fig. 5.5), which is known in the automotive industry as a "black box" (design and cost opaque).

Nevertheless, it is possible to establish the reference cost of the electronic airbag control unit using the precise reference cost calculation model described above; the results of the reference cost calculation are shown in Table 5.6.

The reference costs can be calculated in two ways, either on the basis of a physical prototype or on the basis of a technical design or a technical concept.

5.3 Design to Cost (Design to Cost)

Once the target costs for a product have been determined and broken down into components or subsystems, the components or subsystems are designed according to the target costs. Three methods for design to cost are described below.

Table 5.6 Results of the reference cost calculation for the electronic airbag control unit

Item	Costs (CNY)
Material	
Microcontroller L9689CCP	40.27
Acceleration sensor SAHXC2336A 72F80LR	16.16
Chip ASIC	1.54
Diode	6.16
Electrolytic capacitor	8.05
Chip ASIC 4254X	3.08
CAN driver	1.03
Capacitor	0.31
Chip MOSFET	0.51
Diode	2.57
Transistor	1.23
Operational amplifier	0.21
Pin	9.14
Quartz	1.03
Resistors	0.34
Capacitor	1.68
Printed circuit board	5.39
Connector	2.57
Housing 1	1.54
Housing 2	1.54
Manufacturing	
Applying soldering adhesive	0.08
SMD 1	2.05
SMD 2	1.46
Soldering	0.53
Mounting	0.38
Testing	0.21
Overhead costs	12.48
Reference costs	121.50

5.3.1 Morphological Box

The morphological box is a method for finding global solutions in which a complex subsystem is broken down into relatively independent components, different technical designs are created for each component and a cost calculation is carried out for the technical designs of each component. The best combination of engineering designs to meet the target cost of the subsystem is then selected from the morphological box, which is called the morphological box method.

Below is an example of a morphological box for a left front headlamp (see Fig. 5.6). The headlamp is subdivided into components such as low beam, high beam, blinker, daytime running light, front cover and rear cover. The low beam and high beam are available as halogen, full LED, with costs ranging from low to high. The blinker is also available with halogen, several single LEDs and a single power LED + light guide in three versions and with three different costs. The daytime running light is available in two versions (12 single LEDs, 2 power LEDs + light guide) with different costs, and so on for the front and rear cover designs. Based on this information, the morphological box was listed and the combination of components was selected to match the target cost of the headlamp subsystem:

- Low beam and high beam with design solution halogen lamp (120 CNY),
- Indicators with 7 individual LEDs (45 CNY),

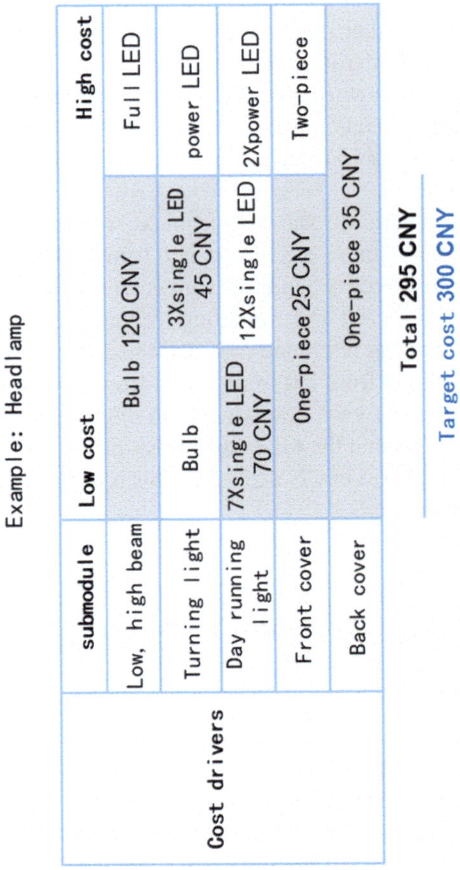

Fig. 5.6 Morphological box for a left headlight

- Daytime running light with 12 individual LEDs (70 CNY),
- One-piece front cover (25 CNY) and
- Rear cover in one piece (35 CNY).

The cost of the whole technical solution of this headlight is 295 CNY, which is in line with the target cost (300 CNY).

The main steps of the morphological box method are as follows.

(1) Decompose the subsystem into relatively independent multiple components.
(2) Create different designs for each component and have a cost engineer calculate the corresponding costs.
(3) For each component, select one of the design options and combine them into an overall design for the subsystem that meets the target cost.
(4) The most important step in the morphological box method is (2), which involves starting from the function/performance requirements of the component and identifying the different designs of function/performance elements to achieve these functions/performances, which is also a litmus test of the engineer's experience and technical competence.

Another practical example is the application of the morphological box method for an interior door panel.

As shown in Fig. 5.7, the interior door panel is first disassembled into the following components: upper panel, middle panel, armrest, lower panel, door pocket, switch panel, speaker cover, trim, inner handle, inner handle speaker cover, inner handle base.

There are various technical solutions and corresponding costs to realize these components, e.g. the upper trim and the middle trim are injection molded and laminated by hand; the trim is laminated and matt chrome plated; etc.

For each corresponding technical design, the cost engineer calculates the corresponding reference cost, and the different technical designs for each component are ordered from left to right and from lowest to highest cost, creating a morphological box. For each component, an engineering design is selected in the morphological box so that the total cost of the interior door panel is CNY 241.45, which meets the target cost of CNY 272.91. This is the practical application of the morphological box method.

5.3.2 Function-Cost Matrix

In Sect. 3.2, when describing the relationship between product function/performance and product value, an example of a function-cost matrix for a quartz watch was shown.

Figure 5.8 shows a function-cost matrix of an instrument cluster for a passenger car. The target cost for this instrument cluster is CNY 520.

According to an analysis using the function-cost matrix method, the 7.5″ TFT display as a function carrier overlaps with the stepper motor, the dial and pointer,

subsystem : AS Door interior trim								target costs (CNY)	272.91
part	**1**	**2**	**3**	**4**	**5**	**6**		**Costs of the selected solution**	
Upper panel	injection molding PP+EPDM-TD15	injection molding PP+EPDM-TD20	positive molded lamination	slush	manual lamination	negative molded lamination	—	**No. 5**	
	20	23	45	90	55		—	**costs (CNY)**	**55.00**
middle panel	injection molding PP+EPDM-TD15	injection molding PP+EPDM-TD20	manual lamination	—	—	—	—	**No. 3**	
	20	23	30	—	—	—	—	**costs (CNY)**	**30.00**
Armrest	manual lamination	—	—	—	—	—	—	**No. 1**	
	35	—	—	—	—	—	—	**costs (CNY)**	**35.00**
Lower panel	injection molding PP+EPDM-TD15	injection molding PP+EPDM-TD20	positive molded lamination	—	—	—	—	**No. 1**	
	50	55	110	—	—	—	—	**costs (CNY)**	**50.00**
Door pocket	injection molding PP+EPDM-TD15	injection molding PP+EPDM-TD20	—	—	—	—	—	**No. 1**	
	8	10	—	—	—	—	—	**Kosten (CNY)**	**8.00**
Control panel	graining ABS	high gloss black ABS	high gloss black PC+ABS	—	—	—	—	**No. 3**	
	6	20	25	—	—	—	—	**Kosten (CNY)**	**6.00**
Loudspeaker cover	injection molding one-piece	injection molding+high gloss black deco-ring	Injection molding+matt deco-ring	—	—	—	—	**No. 1**	
	10	20	25	—	—	—	—	**costs (CNY)**	**10.00**
Trim strip	matt black	high gloss black	IMD	matt black+high-gloss chrome-plated	high-gloss black+high-gloss chrome-plated	IMD+high-gloss chrome-plated	lamination+ matt chrome-plated	**No. 7**	
	15	25	35	30	40	50	35	**costs (CNY)**	**35.00**
Innengriff	without painting	graining	high gloss painted	matt painted	matt chrome-plated	hight gloss chrome-plated	—	**No. 5**	
	1.35	1.3	—	5.85	8	6.5	—	**costs (CNY)**	**8.00**
Speaker cover on the inside handle	without painting	graining	high gloss painted	matt painted	matt chrome-plated	hight gloss chrome-plated	—	**No. 2**	
	1	0.95	13.8	6.5	7.2	5.7	—	**costs (CNY)**	**0.95**
Inside handle base	without painting	graining	high gloss painted	matt painted	matt chrome-plated	hight gloss chrome-plated	—	**No. 2**	
	—	3.5	—	8.6	—	—	—	**costs (CNY)**	**3.50**
total costs for the selected solution combination (CNY)									**241.45**

Fig. 5.7 Morphological box of an interior door panel

function-cost matrix (example: instrument cluster)						
function category	main function		secondary function		image function	
carrier / function	display vehicle velocity, engine rotation rate and other signals (CNY)	warning display (braking fluid, airbag etc.) (CNY)	visibility (sun shield) (CNY)	protection & parts fasten (CNY)	grant image (CNY)	total function cost (CNY)
PCB incl. EE components (CNY)	112.50	37.50				
step motor (CNY)	20.00					
instrument dial & cursor (CNY)	7.50	7.50				
7.5inch TFT display (CNY)	87.50	87.50			175.00	
LED (CNY)		10.00				
front housing (CNY)			3.75	1.25		
decor strip (CNY)					3.00	
back cover (CNY)				3.50		
connectors & wiring harness (CNY)	2.50	2.50				
fasteners (CNY)				3.00		
total (CNY)	**230.00**	**145.00**	**3.75**	**7.75**	**178.00**	**564.50**
total (%)	41%	26%	1%	1%	32%	100%

Fig. 5.8 Function-cost matrix of an instrument cluster for a passenger car

and the LEDs of the warning functions. In addition, if the design exceeds the target cost, the trim, which is the function carrier for the image function, can be omitted, reducing the cost by 3 CNY.

The 7.5-inch TFT display can be reduced to 3.5-inch to reduce the cost by 50 CNY, so that the final cost of the instrument cluster after optimization by the function-cost matrix method is 513.50 CNY, which can then meet the target cost of 520 CNY.

The elements of the function-cost-matrix approach are.

Describe clearly and concisely,

- What the primary functions of the subsystem are.
- What are the secondary functions?
- What are the image functions?
- Is there more than one function carrier for the same function?
- Can the same function be realized with a less costly function carrier?
- Are all image functions required? Which ones can be eliminated in terms of customers and competition?

With these standardized questions, it is relatively easy to find your way around the function-cost matrix and work towards the target costs.

5.3.3 *Benchmark Function*

For a subsystem or a component, different brands and products on the market usually have different technical designs and correspondingly different costs. The benchmark function method is based on the analysis of competitive products and functions in order to create a design at the target costs.

Take the example of a car wing mirror:

there are five car models on the market with different exterior mirror designs at different costs. These five exterior mirrors cost between 131 and 243 CNY. The essence of the benchmark function method is to first select the most cost-effective design from these five exterior mirrors, which include the basic functions, and derive the final design based on a waterfall diagram:

- Continuous improvement process (CIP) over the entire product life cycle including the effects of inflation
- Reduction of the purchase price by increasing the purchasing volume
- Additional functions (e.g. LED) on the target exterior mirror compared to the exterior mirror of the benchmark (vehicle model 3)
- The final result is the cost achieved by analyzing the benchmark function method in the amount of 166 CNY, thus conforming to the target cost of 170 CNY (see Fig. 5.9).

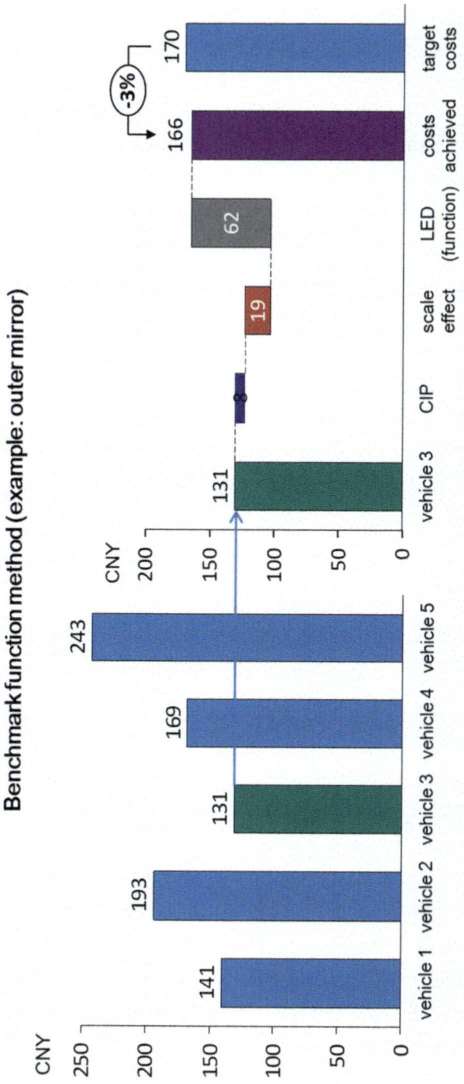

Fig. 5.9 Application of the benchmark function method for an outer mirror

5.4 Value Analysis and Value Engineering (VAVE) of Components and Assemblies

VAVE requires a cross-divisional, interdisciplinary team of employees (see Fig. 5.10), whether from R&D, costs, quality, production or purchasing, sales, finance or project, who all have the customer in mind in order to carry out a VAVE for each component and each assembly (subsystem).

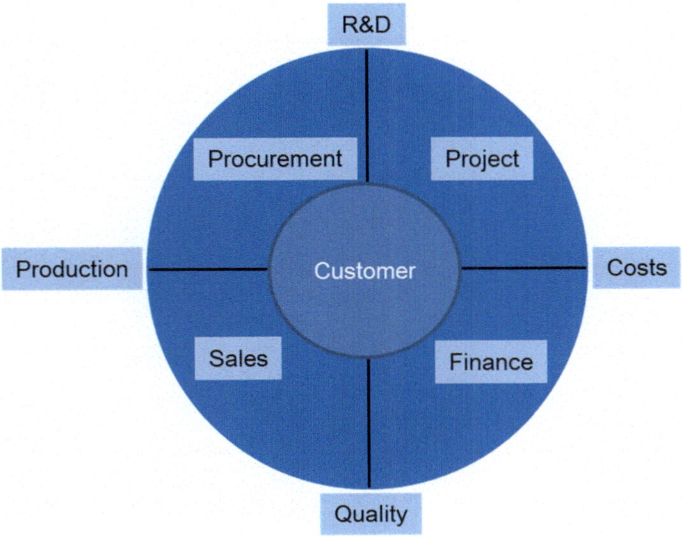

Fig. 5.10 Cross-divisional and interdisciplinary team for value analysis

5.4.1 Value Analysis

The value analysis of components and assemblies (subsystems) is mainly used in the series production phase of the product and is divided into six areas.

- Function
- Material
- Manufacturing
- Tolerances and standards
- Surface treatment
- In-house production or purchasing (make-or-buy).

The steering knuckle (see Fig. 5.11) is a typical component with functional/performance requirements, the corresponding static and dynamic loads, materials used, manufacturing processes and tolerances, surface treatment and make-or-buy characteristics.

How do you carry out a value analysis of components and assemblies?

Based on more than 25 years of experience in vehicle development and cost engineering, the "standardization questionnaire-guided value analysis" is presented here, which enables the value analysis team to systematically and comprehensively analyze and evaluate components or assemblies in order to find innovative and optimized design solutions.

Below are some typical standardization questions, and there are many more in-depth standardization questions for a value analysis workshop, which help the value

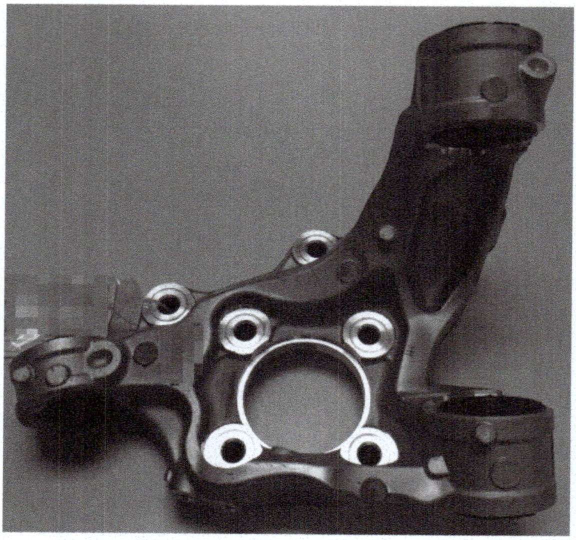

Fig. 5.11 Left rear steering knuckle of an SUV

analysis team to analyze and investigate the technical cost reduction potential of the component and assembly in all technical and commercial aspects.

The following standardization questions are examples of the six areas mentioned above:

1. Function

   (1) What are the main functions that the component must fulfill?
   (2) What secondary functions must the component fulfill?
   (3) Are all functions designed to meet the intended objectives?
   (4) Can some of these functions be fulfilled with another, already existing component?
   (5) How does the component work?
   (6) What are the individual functional parameters of the component?
   (7) Are the requirements for the component higher than the normal operating range?
   (8) Are these requirements compatible with the quality requirements of the assembly?
   (9) Are these requirements higher than those of comparable competitor products?
   (10) Can the number of uses be reduced if the component is used several times in the same vehicle?
   (11) Are these functions important for most customers?
   (12) etc.

For example, if there is a metal bracket in the engine compartment that holds the wiring harness in place, and another bracket for the air condition pipe in the engine compartment that also needs to hold in place, standardization question #4 could be used:

> Can some of these functions (holding a wiring harness in place) be accomplished with another existing component (bracket for the air condition pipe)?

Could the bracket that holds the air condition pipe be modified (by punching a hole or "adding" an "ear" during punching) to hold the wiring harness in place? If the local arrangement is feasible, this would fulfill the fixing function for the air condition pipe, and a separate holder is eliminated, reducing the cost by 10 CNY and the weight by 250 g.

2. Material

   (1) Can another material be used?
   (2) Does this material meet the requirements of the company standard?
   (3) Does the material meet the requirements of national (GB) or international standards (ISO)?
   (4) Is it possible to use a less expensive material?
   (5) Which dimensions influence the material weight of the component?
   (6) Are the design dimensions of this component too large?
      a. Have the design dimensions of the part been analyzed with CAE?
      b. Are the components of the assembly oversized?
      c. Are the dimensions too large compared to competitors?
      d. Have the dimensions of the components and assemblies been verified by testing?
   (7) Can some of these dimensions be reduced?
   (8) What percentage of the weight is material utilization?
      a. Can the material utilization ratio be increased through design or manufacturing optimization?
      b. Can the material waste be reused and for what purpose?

(9) etc.

Take the example of a loudspeaker trim ring (see Fig. 5.12) in its original state: material ABS; surface treatment: matt chrome-plated.

The new surface coating process "painting with aluminum paint" offers the same visual and haptic effect and the same surface resistance as matt chrome plating, and the substrate material can be changed from the more expensive ABS to the more cost-effective PP. By optimizing the material and the surface coating process, the cost of this decorative ring could be reduced from 3.50 CNY to 2.10 CNY.

3. Manufacturing

   (1) How is the component manufactured?
   (2) Can the component be produced using a different manufacturing process?
   (3) Can the components be processed as an assembly?

Fig. 5.12 Loudspeaker trim ring for the left front door panel of a passenger car

(4) Can the assembly be processed as a single component?
(5) How are the components assembled into an assembly?
(6) Can the assembly process be simplified?
(7) Can the assembly time be shortened or can assembly steps in the assembly process be eliminated?
  a. What engineering changes are required to achieve the above objectives?
  b. Or by changing the delivery condition?
(8) Can the components be sourced from specialized suppliers?
(9) etc.

Question (2) "Can the component be produced using a different manufacturing process?" A practical example is the rear left and right steering knuckle (see Fig. 5.13).

Before cost optimization: forging manufacturing process; material forged aluminium 6082, tensile strength 340 MPa, yield strength 300 MPa; load: maximum stress for reverse brake operation 89 MPa.

After cost optimization: Die casting manufacturing process; material cast aluminium ZL101A, tensile strength 295 MPa, yield strength 220 MPa. Considering the maximum stress of 89 MPa in normal driving condition, the component can also be produced by die casting with the material cast aluminum ZL101A, and unit cost reduction of 87 CNY.

4. Tolerances, standards.

(1) Which tolerances increase the cost of components?
(2) Which tolerances can be relaxed a little?
(3) Are similar components in use?
(4) Are they standard parts?

Fig. 5.13 Rear left and right steering knuckle

  a. Is it a national or international standard?
  b. Is it a company standard?
(5) Are standard parts with similar dimensions and designs already in use?
(6) Can this component be replaced by a standard part?
(7) etc.

Question (2) "Which tolerances can be relaxed?" This standardization question was raised in the value analysis team, and after detailed discussion and random sampling, the conclusion was reached:

- that the end face roughness Ra 1.6 μm of the setting arm (see Fig. 5.14) could be reduced to 3.2 μm,
- that the uniform roundness of 0.004 mm for the two round holes could be changed to 0.006 mm roundness for the 9 mm diameter hole, and to 0.009 mm roundness for the 27 mm diameter hole,
- and that the perpendicularity of the inner surface of the round holes can be curled from 0.012 to 0.03 mm,
- thus eliminating some processing steps and reducing manufacturing costs by CNY 7.48.

5. Surface treatment

   (1) What chemical and physical environments affect this component?
   (2) What surface treatments are planned?

Fig. 5.14 Adjusting arm of a motor

(3) Which of these surface treatments are required?
(4) Which ones can be replaced with other surface treatment processes?
(5) Is it possible to apply surface treatments to only part of this component?
(6) etc.

Question (3) "Which of these surface treatments are necessary?" In the sales, quality, cost and R&D departments, there was a heated debate as to whether or not the instrument panel reinforcement beam (see Fig. 5.15) should be electrophoretically coated. Comparing the vehicles of Mercedes-Benz, BMW, Toyota, etc., their instrument panel reinforcement beams are largely not electrophoretically coated because the instrument panel reinforcement beams are mounted in a relatively dry place and also some rust spots do not affect their functions/performances (rigidity, safety, fatigue strength, NVH).

The elimination of electrophoretic coating not only reduces the cost of one piece by 20 CNY, but is also an important step towards environmentally friendly production.

6. Make or buy

(1) In what batch and in what time frame will this component be delivered?
(2) Do our order and the delivery quantity per batch match the actual economic production method (and the actual production efficiency)?
(3) Which other suppliers can deliver the same component and also guarantee the required quality and time frame?
   a. Do these other suppliers already supply other components to us (including subsidiaries)?
   b. Can these other suppliers offer a better price for this component?
(4) Are the current suppliers competitive in terms of production processes or equipment?
(5) Are there similar components already being sourced from other suppliers?
(6) Are the costs for these components comparable?
(7) Can the current supplier generate cost reduction proposals?

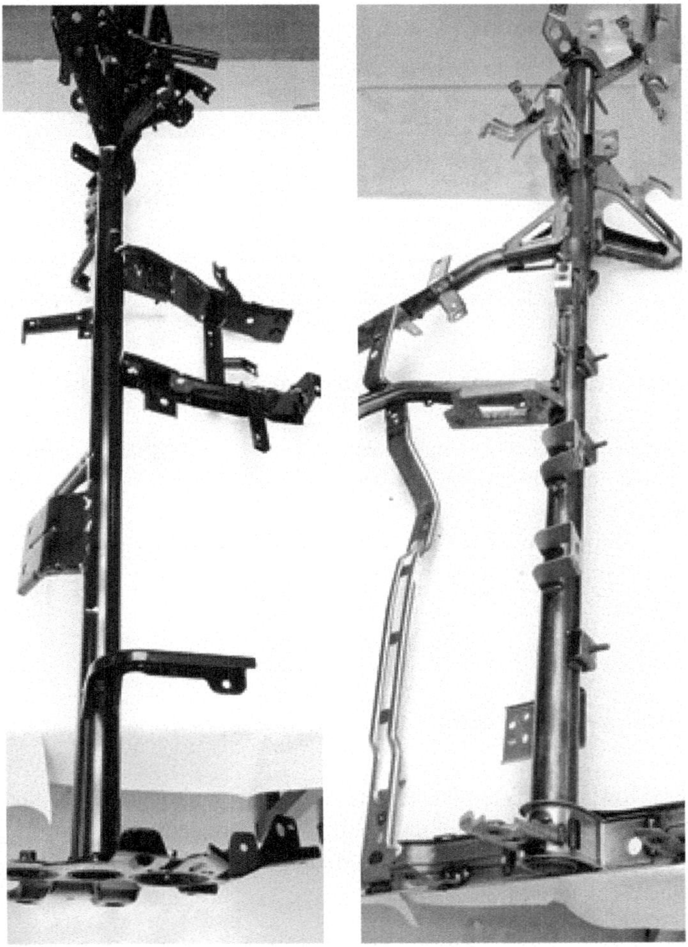

Fig. 5.15 Reinforcement carrier of the instrument panel of a passenger car: Left component with electrophoretic coating (black), Right-hand component without electrophoretic coating

(8) What can be done to reduce the cost of packaging, transportation and storage of the component?
(9) Can this component be manufactured in-house to reduce costs?
(10) Can this component be purchased to reduce costs?
(11) etc.

Question (5) "What can be done to reduce the costs of packaging, transportation and storage of the component?".

Here is an example of screw bolts on the assembly line (see Fig. 5.16):

The supplier uses the returnable plastic boxes with screw bolts directly for the assembly line instead of screw bolts in packing cartons, saving the cost of cartons and

Fig. 5.16 Packaging carton and reusable plastic box of screw bolts for the assembly line

the labor cost of redistribution to the returnable plastic boxes, which is equivalent to a cost reduction of about 300 CNY per day or 72,000 CNY per year.

Value analysis means analyzing the whole process of the product from raw materials, processing, assembly to the finished product to find the optimization potential in every small link.

Continuous improvement is a daily business that must be done, do not settle for the status quo, "there is no best product design, only there will be better product design tomorrow".

5.4.2 *Value Engineering*

Value engineering is mainly applied in the product development phase, where approximately 85% of product costs are incurred, so the application of value engineering in the product development phase is crucial. Two main areas of work form the main tasks of value engineering:

- Design to target costs (design to cost)
- Analysis of functions/performance and costs, for which the methods of function-cost matrix, benchmark function are frequently used

The entire process applies the methods presented in Sects. 5.1, 5.2 and 5.3 and begins with cost optimization in the early design phase, while avoiding unnecessary expenditure on research and development and tool changes.

5.5 Linear Regression Analysis

Linear regression analysis is suitable if a component or a subsystem has to perform certain functions/performances and if one of the functions/performances is dominant and the costs of the component or subsystem are mainly determined by a functional

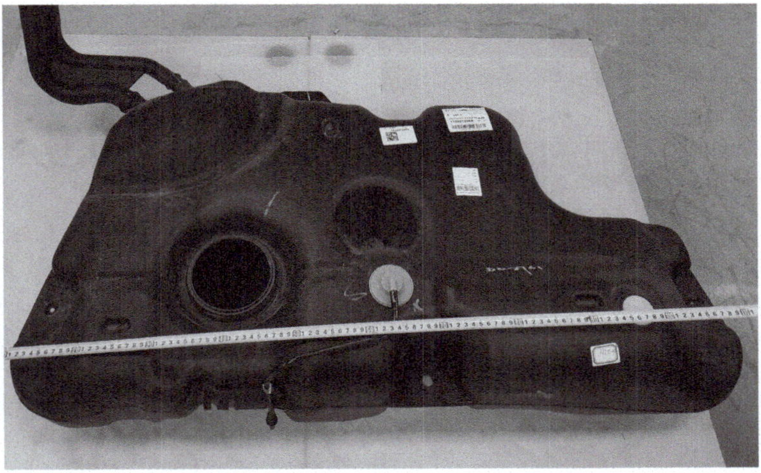

Fig. 5.17 Fuel tank of a passenger car

parameter of this function/performance. Let's start with an example: the fuel tank of a passenger car (see Fig. 5.17).

The most important functional parameter of a fuel tank is its volume. Figure 5.18 shows the price distribution of four fuel tanks with different volumes, here with four data sets (volume (x) liters, price (y) CNY): (45, 280); (50, 320); (60, 410); (75, 560).

Linear regression analysis is a method of statistical mathematics that assumes that the influencing variable is x and the target variable is y. Linear regression analysis consists of finding a linear function as follows.

$$y = ax + b \tag{5.2}$$

Fig. 5.18 Results of the linear regression analysis for fuel tanks

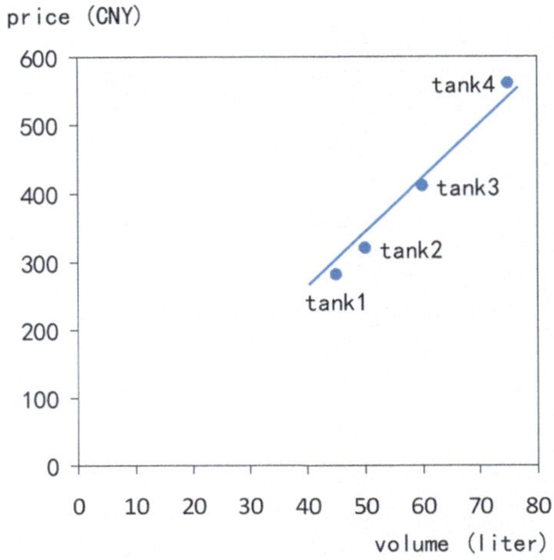

whereby

$$y_i = a_i x_i + b_i + \epsilon_i, \quad i = 1, 2, 3, \ldots, n \tag{5.3}$$

and ϵ_i is the error of the linear regression function in relation to the ith data set.

The linear regression function for the four data sets mentioned above is

$$y = 9.33x - 139.85 \tag{5.4}$$

If you want to know the approximate price of a 70-L fuel tank, insert x = 70 into the above regression function (5.5), the price of a 70-L fuel tank is CNY 513.30 (see Fig. 5.18).

Batteries for an electric vehicle are also well suited for the application of linear regression analysis, where the x-variable is the battery capacity (kWh) and the y-variable is the price. Here, the battery capacity and the corresponding price of several batteries are entered, resulting in the following linear regression line.

$$y = 751.84x - 995.37 \tag{5.5}$$

If we want to know the price of a battery pack with a capacity of 90 kWh, the linear regression function results in a price for the battery pack of about 66,670 CNY (see Fig. 5.19).

The linear regression analysis method is well suited to situations where the key functional/performance parameters of the component or subsystem are paramount and have a major impact on cost. The price (cost) of the component/subsystem can be calculated relatively quickly by applying the linear regression method.

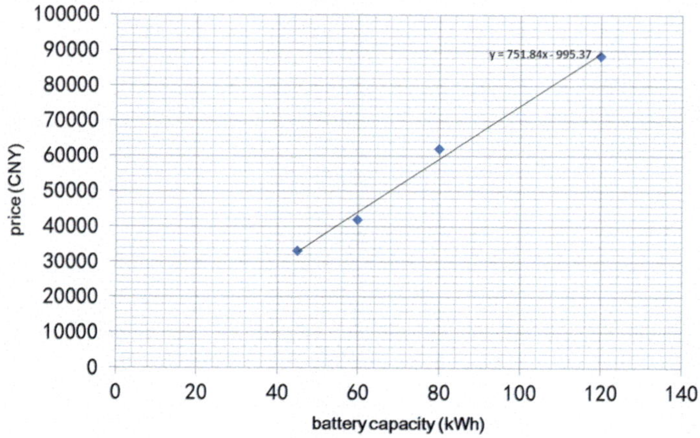

Fig. 5.19 Results of the linear regression analysis of batteries for electric vehicles

Table 5.7 Product design for an interior door panel

Part	Design 1	Design 2	Design 3	Benchmark design
Base body inner lining (CNY)	125,50	120,20	131,79	120,20
Protective film (CNY)	12,05	10,65	13,68	10,65
Window regulator control panel (CNY)	45,02	51,46	55,35	45,02
Trim strip (CNY)	15,66	10,14	16,75	10,14

Table 5.8 Manufacturing steps and processes for an interior door panel

Manufacturing steps and processes	Design 1	Design 2	Design 3	Benchmark design
Injection molding (CNY)	17.50	18.25	15.80	15.80
Assembly (CNY)	4.55	5.25	4.05	4.05
Total costs (CNY)	220.28	215.95	237.42	205.86

5.6 Best of Bench

For one and the same product, there may be different product designs and correspondingly different manufacturing steps and processes. The best-of-bench method means combining the most cost-effective product design with the most efficient manufacturing steps and processes to obtain the best solution for the product (best of bench).

The first step is to list the different designs of a product, here an example for the interior door panel of a passenger car. The product design, manufacturing steps and processes are shown in Tables 5.7 and 5.8.

What are the main differences between the three designs? Design 2 has the advantage that the inner body is 2.5 mm thick (3.0 mm thick for designs 1 and 3).

In the manufacturing process step "injection molding", design 3 has a faster manufacturing time of about 60 s than design 1 and 2 (difference in the production plant and in the process parameters), and design 3 is also the best in the manufacturing process step "assembly".

Ultimately, the optimum cost is CNY 205.83, which none of the three designs can achieve on its own and which should be used as a cost limit for future improvements.

5.7 Portfolio Analysis Method

This refers to the analysis of technology/product segments rather than the analysis of investment target segments commonly used in the capital markets. The steps in conducting a technology/product segment analysis are.

- Identify a technology/product
- Analyze the current status of the technology/product
- Predict the future status of the technology/product
- Analyze the function/performance of the technology/product and the two-dimensional distribution of function carriers/performance carriers.

Figure 5.20 shows the portfolio analysis conducted by CostKey Solutions (Shanghai) Co. Ltd. for electric SUVs in the Chinese market in 2019. The average range (NEDC) in 2019 is 348 km and most electric SUVs are expected to reach 500 km (or more) in 2022, with the required battery capacity increasing from an average of 51–75 kWh.

5.8 Analytical Functional Analysis of Manufacturing Costs and Production Volume

The manufacturing costs account for about 10–15% of the product costs, part of which is accounted for by the depreciation costs of the production line. Assuming that the depreciation costs of the production line are a function f, the analytical functional relationship between the function f and the production volume n is.

$$f(n) \sim I\frac{1}{n(1+i)^J}, \quad n \le N \tag{5.6}$$

$$f(n) \sim 2I\frac{1}{n(1+i)^J}, \quad N \le n \le 2N \tag{5.7}$$

where i is the annual return on investment, I is the amount of investment in the production line, J is the number of years of depreciation for the production line and N is the number of units at full production capacity of a production line. If n is greater than N, the depreciation for the production line increases sharply as the investment in another production line becomes necessary (see Fig. 5.21), and so on.

Taking an injection molding line for the front bumper of a passenger car with an investment I of CNY 12 million, a depreciation period J of 15 years and an annual return on investment i of 7.5% as an example, the calculated annual depreciation cost of the line is CNY 0.80 million. If the production volume increases from 100,000 units p.a. to 150,000 units p.a., this results in a cost reduction of 7% (see Table 5.9).

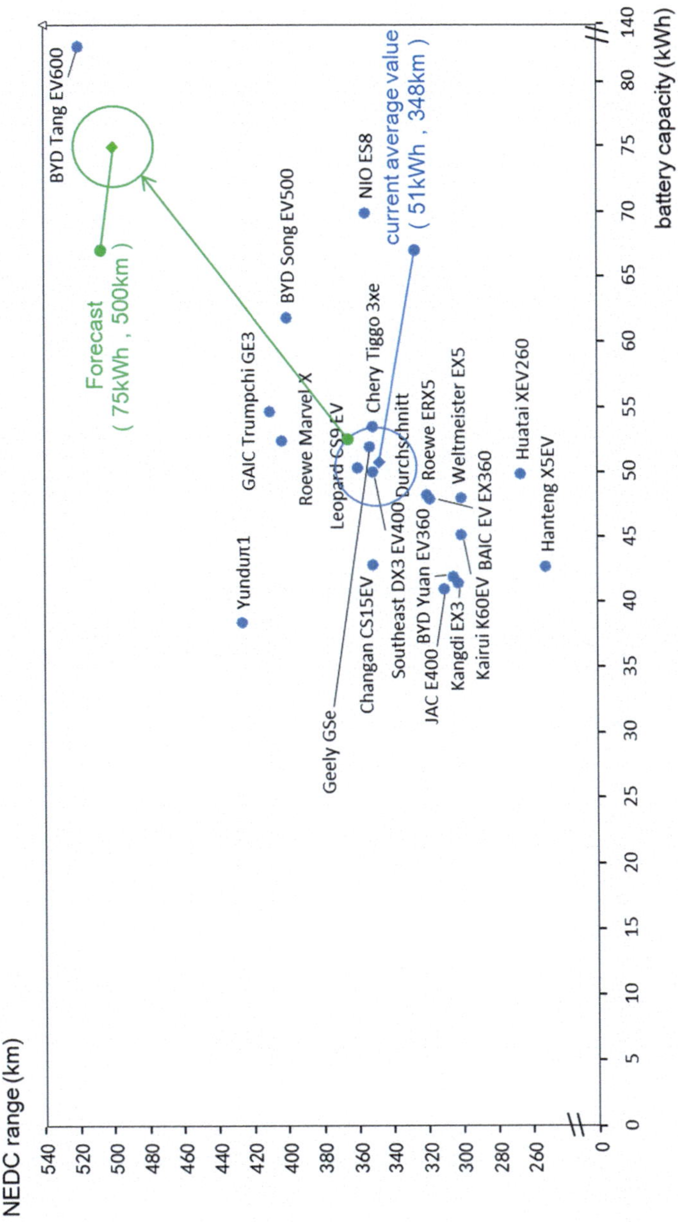

Fig. 5.20 Portfolio analysis for electric SUVs on the Chinese market

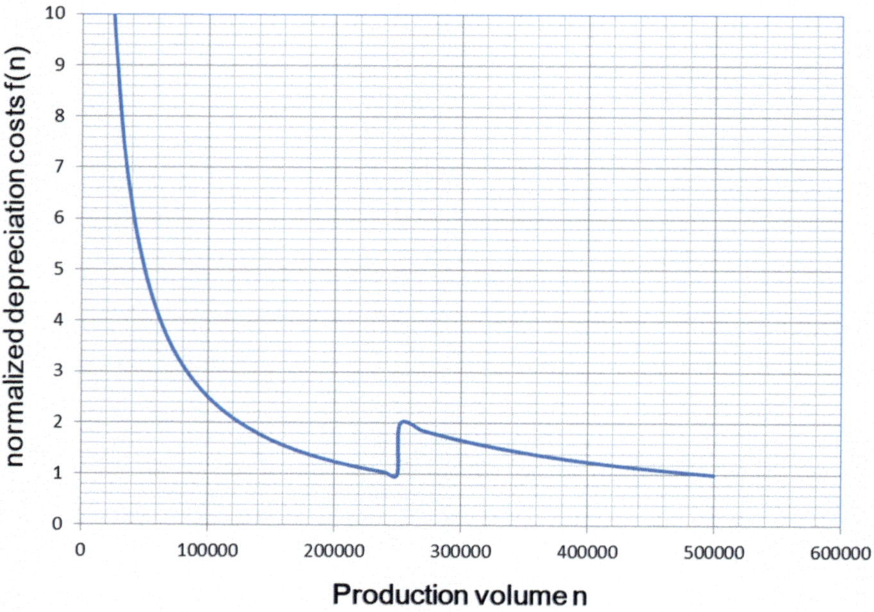

Fig. 5.21 Normalized depreciation costs of the production line (N is 250,000)

Table 5.9 Production volume and cost difference

	Production volume 100.000 units p.a	Production volume 150.000 units p.a
Material costs	65.52	65.52
Production costs	22.00	15.83
Subtotal	87.52	81.35
Cost difference (based on 100,000 units p.a.)	–	– 7%

5.9 Business Negotiation Strategy and Tactics

The reference costs of a product and cost breakdowns calculated by the cost engineers are exactly what are needed in the business negotiation of product prices. In some advanced manufacturing companies, the cost engineers are involved in and even lead the business negotiations with the suppliers, which is a practical realization of the above approach.

In his Daimler days, the author has participated in or led hundreds of business negotiations, including, for example, a four-month successful business negotiation with a major supplier on the ESP subsystem, which was described as an in-house best practice.

Cost engineers need to master the art of good negotiation. The cost engineer's understanding of product design and familiarity with production processes also make them a productive negotiating partner.

5.9.1 *Negotiation Goals and Motivations*

Business negotiations usually involve a product supplier on one side and a product buyer on the other, with the product supplier aiming for a high product price and the product buyer aiming for a low product price.

The product supplier is motivated by the desire to achieve high profits at a high product price; the product buyer is motivated by the desire to purchase at a low product price in order to reduce the cost of supplier parts and increase the company's profitability.

Both parties have the same goal and motivation, but the outcome of the negotiation is opposite. A clear understanding of these two points is at the heart of business negotiations and is the starting point for both parties to apply the different negotiation strategies and tactics.

5.9.2 *Preparing for the Negotiation*

Do not go into business negotiations unprepared; it is important to be prepared. Figure 5.22 helps the reader to recognize what to prepare for before a negotiation.

1. Target

First of all, the target of the negotiation must be clarified before the business negotiation: Does the buyer of the product want to reduce the purchase price by 3%? Does the supplier want to increase the price by 2% because the customer has made a engineering change?

2. Information

(1) The current purchase price of the product?
(2) Breakdown of the costs of the product: Breakdown of material costs, breakdown of manufacturing costs, breakdown of administrative costs (overhead)
(3) What is the purchase volume of the product? What is the total amount?
(4) What is the quality ratio?
(5) What is the market penetration (how many manufacturers can offer the same or similar products)?
(6) What are the operating conditions of the negotiating partners (turnover, market share, profitability, future development prospects)?

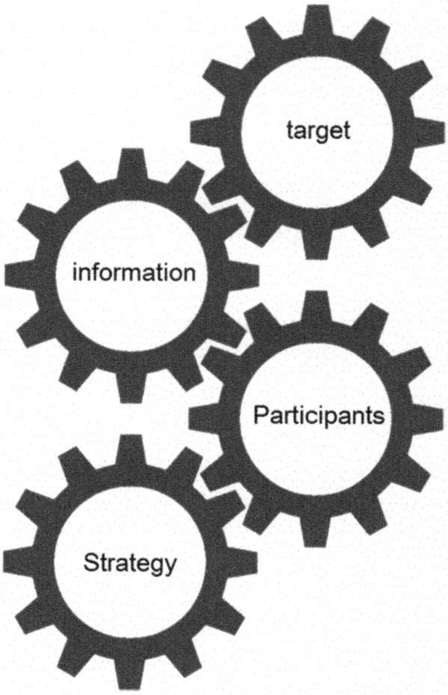

Fig. 5.22 Elements of preparation before a business negotiation

3. Participants

Who is involved in the negotiation, who is the lead negotiator on one side, what is the expertise, style of the lead negotiator?

4. Strategy

Once the negotiation target has been clarified, the negotiation strategy is the path to the negotiation target and the implementation measures within this path. It is important to consider whether the negotiation is a win–win strategy or a zero-sum game.

If it is a zero-sum game, the party that wins the negotiation today may lose the customer or product supplier tomorrow. It is the win–win strategy of negotiation that helps your own company to achieve sustainable development.

5.9.3 The Tactics of the Negotiation Process and the Art of Negotiation

Negotiation means:

- Discuss, act, negotiate
- Convincing the other party with the advantages and disadvantages
- Rhetoric, emotional intelligence, charm, commitment
- Winning over your opponent

- Negotiate fairly.

Business negotiations start with arguments and require arguments, e.g. what do the two sides have in common when it comes to the cost breakdown? What are the differences? Why are there these differences?

Sometimes an agreement is also necessary. What leverage do you have that you can give the other side in exchange for a lower price? In most cases, a fair negotiation process and outcome is required. Serving tea and coffee to your guests on the negotiating side is hospitality and is one of the basic skills of a good negotiator.

Small talk is very important at the first meeting and influences the atmosphere and the course of the negotiation. Give your negotiating partner the opportunity to reflect on the situation and give them the chance to report back to their superior or ask for advice. Set up an internal meeting room where the negotiating team can communicate with the guests.

It is not just one negotiation meeting that is necessary or that leads straight to the goal, but several that take place at a certain rhythm. Sometimes a compromise is necessary (Table 5.10).

Table 5.10 Matrix of the art of negotiation

The art of communication	Game	Participants	Style	Appearance image
• Opening, negotiating • Offensive and defensive communication • Organization of the topics • Order of discussion topics • Escalation strategy	• Sequence of arguments • Techniques of argumentation • Techniques of contradiction • Argumentation chain • Compromise plan • Negotiation jeton	• Negotiator • Negotiation participants • Roles and tasks • Alternation of the negotiator • Seating arrangements	• Polite • Credible • friendly • trustworthy • fair • aggressive	• "Hard", "soft" presence • Greeting, politeness, small talk • Business rules • Clothing • Body language • Corporate impression
Techniques	Methods	Time	Organization	Negotiating partner
• "Good Guy, Bad Guy • "Old school" techniques • "Control" • "Threats, intimidation"	• "Salami tactics" • Resistance • Playing for time • "Give and take" • Interruption	• Appointment agreement • Time • Time pressure	• Location • Reception of guests • Guest transportation, accommodation	• Target • Arguments • Strategy • Manner

5.9.4 Concluding Negotiations and Summaries

It is a good negotiation result if the negotiating parties can agree on a win–win price range (see Fig. 5.23).

Just like the preparation before the negotiation, the negotiation must also be summarized after the result has been achieved:

- Whether the outcome of the negotiation was in line with the intended target
- Whether the result was within the win–win price range
- Which negotiation jetons were exchanged?
- What points in the negotiation process need to be improved in the future?

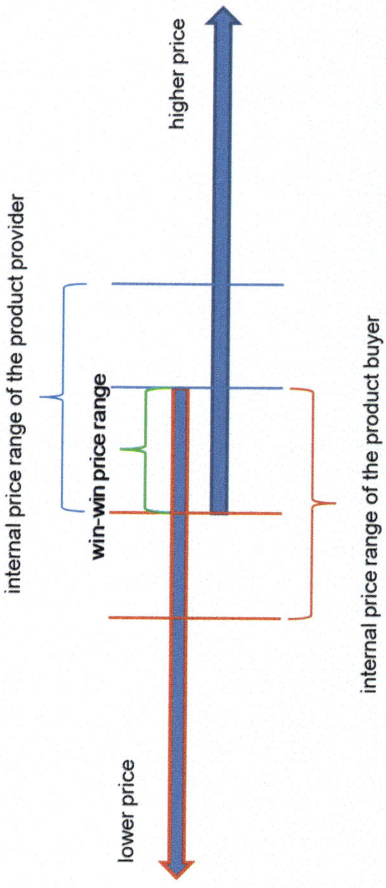

Fig. 5.23 Win–win price range

In short, business negotiation is strategic playing of wisdom, market and technology, not only pursuing short-term goals, but medium and long-term strategies and interests are the core of business negotiation. Business negotiations are not a zero-sum game, but win–win situations that can lead to the sustainable development of your company.

Here are some examples of negotiations that the author has personally experienced:

(1) Price negotiation of a forged part in a gear shift

The supplier was a small production company on Lake Boden in Germany; the customer was the Mercedes-Benz transmission plant in Gaggenau; the quoted price for the forged part was €0.35; the reference costs determined by the cost engineering department were €0.25.

The negotiations took place at the supplier's production facility and were attended by the managing director of the supplier, the head of the production line, the author himself and an employee of the Mercedes-Benz cost engineering department.

First, we were given a tour of the production line for forged parts accompanied by the supplier's managing director, who gave us a detailed explanation of the production steps.

The negotiations took place in the conference room, where the cost engineering department employee first presented the premises, cost calculation methodology and details of the reference cost calculation for this forging.

After the presentation of the reference costs, the supplier's managing director said: "If we had gotten a price of 0.25 EUR, we would have had to close the plant!" This was a very small manufacturing company that accounted for more than 50% of the supplier's turnover with orders from Mercedes-Benz and had been supplying Mercedes-Benz for many years without any quality problems and with good supply reliability.

At this point, there was really nothing more the author could do and he thought that if it was a large, dominant supplier, he would have to negotiate with it to the bitter end. In addition, the forged part was a key component for which there could be no quality or logistics errors. So the author said to the supplier's managing director: "Well, we've seen the production line, thanks for the explanation, and when we get back to Gaggenau, we'll coordinate internally and I'll call you back."

Two days later, after an internal vote, we decided on a compromise: if €0.25 was not achievable, €0.30 was feasible. In the end, the author and the supplier's managing director agreed on €0.29 over the phone.

Notes on this:

The supplier

- The supplier was cooperative, open and transparent throughout the negotiation process

- The managing director was personally present and presented the relevant production lines for forged parts, which shows the importance attached to the discussion with the customer.
- Transparent company reality and good customer reputation

The customer (Mercedes-Benz)

- Well prepared and pre-calculated reference costs of the component
- Transparent presentation of the reference cost breakdown
- Appreciation of the supplier's goodwill and opportunity to inspect the relevant production line
- Interruption of the negotiation rhythm to give the supplier and ourselves the opportunity to maneuver (face-to-face negotiation followed by a phone call to reach a consensus)

The end result of the negotiations was a win–win situation.

(2) Tooling costs for the steering column switching module.

The supplier is a large subcontractor based in the northern suburbs of Stuttgart, Germany; the tooling cost for the steering column switching module was quoted at 1.9 million EUR; the reference tooling cost calculated by the Cost Engineering department was 760,000 EUR.

The negotiations took place in a meeting room at the customer's side and were attended by the supplier's sales manager, the author himself and a cost engineer responsible for tooling costs. The difference between the quoted tooling costs and the reference costs was so great that the author decided to conduct this negotiation in a "cold way".

After exchanging pleasantries, the customer indicated to the supplier's sales manager that there were coffee and butter pretzels on the table, to which the author said, "The quoted tooling cost of 1.9 million EUR is very high, we will send you a breakdown form of the tooling cost, please fill it out and send it back to us in four weeks", "This is a standard process with us". The supplier's sales manager nodded in agreement, drank his coffee, ate a pretzel, and then took his leave. The negotiation session lasted about 30 min in total.

Four weeks later, the supplier's sales manager sent a detail of the tool and offered €900,000 for it. The other €1 million "disappeared". Compared to the reference tool cost of €760,000 calculated by the cost engineer, the author decided "that's fine". An agreement of €900,000 for the cost of the tool was reached in the manner of "no blood on the sword".

Notes on this:

- It is not uncommon for suppliers to overstate their bids.
- "No blood on the sword" is a fine art of negotiation

(3) Electronic control unit ESP.

The supplier was a large supplier ("Fortune 500"); the supplier offered 147 EUR; the reference cost was 125 EUR. The negotiations took place in a conference room and were attended by the sales manager, the key account manager and the technical development representative of the supplier, as well as the purchasing manager, the cost engineer for electronics and the author himself on the customer side.

After the personal introduction of both sides, the cost engineering employee first presented the three components of the reference costs: material costs, production costs and administrative costs, and asked the supplier for corresponding comparative data. The supplier's sales manager replied: "We have no such precedent in our company and cannot provide any data on these three cost components".

The author had anticipated such a reaction and said, "I tell you what, we can't negotiate at this level, please escalate to your director level and we will do the same", "How much time do you need for an answer?" The supplier said, "In a month". So it was agreed that we would meet again in a month.

A month passes, and the cost engineering employee asks the supplier's key account manager: "No answer from senior management yet". Almost two months later, the cost engineering employee asks the supplier's key account manager again and he replies: "Oh, senior management has agreed that we should give you the three-part cost components, but no further breakdown!" This was a big step forward, and two months later the parties were back at the negotiating table with the same people in the conference room.

At this time, the author asked the cost engineering person to present the reference cost breakdowns for the three cost items and said, "Please look at these breakdowns and point out where you agree and disagree." The negotiations were still as tough and tasteless as a long chewed piece of gum.

Gradually, however, the author realized that there were major differences between the two sides on several cost breakdowns. At this point, the author asked the supplier for a tour of the production line, and to the customer's surprise, the supplier's key account manager readily agreed. The production line was located in the Black Forest, and the colleagues and the author were given a closer look at the line, plant equipment, production cycle, etc. used to manufacture the ESP electronic modules.

Four months passed, and after this "protracted" negotiation, the two parties finally agreed to reduce the supplier's offer by 11.6 EUR (– 8%).

Notes:

The supplier was a "large supplier in excess". The customer was also no joke, but the author's colleagues and managers were not optimistic about negotiating with the "Goliath" at the time.

A "protracted" cat and mouse game

Persistence and perseverance are one of the keys to negotiation, and escalation to management is also a form of negotiation. Communicating the cost breakdown requires patience and transparency on both sides. The final result is, against all odds, a win–win negotiation.

5.10 Business Case Calculation (Net Present Value: NPV)

If a product is to be developed, produced and launched on the market, a net present value (NPV) calculation is required. The value of the NPV can intuitively be understood as the total net profit that a product can bring to a manufacturing company during the entire product life cycle, from the first day of product definition to the last day of series production.

The following information is required for the business case calculation.

- Expected annual sales volume of the product during its entire life cycle
- Expected sales price of the product and it's change in the course of time
- One-time costs (R&D costs, tool costs, costs for production facilities)
- Sales, administration and other costs
- Target profit margin
- Interest rate on capital
- Tax rate.

The following is an example of a business case calculation (NPV) for an electric car (see Fig. 5.24). Based on the forecast sales volumes and sales prices in 2018, 2019, 2020 and 2021, R&D costs of CNY 200 million, a capital interest rate of 8%, a target profit margin of 6.5% and a tax rate of 13%, the NPV over the life of the product is calculated at CNY 391 million.

The business case calculation also includes a sensitivity analysis of the NPV, with the most important sensitivity being the question of how much the NPV changes with a 5% change in product costs. There is also the question of how much these would affect the NPV if the sales volume were to change by 5% and the sales prices by 5%.

As shown in Fig. 5.24. A 5% increase in costs above the target costs reduces the NPV from 391 to 193 million.

A 5% decrease in selling price reduces NPV to CNY 138 million.

If the sales volume decreases by 5%, the NPV decreases to 362 million NPV.

This shows that the NPV of a product over its entire life cycle is most dependent on the cost and selling price, which means that compliance with.

Fig. 5.24 Business case calculation and NPV sensitivity analysis for an electric car

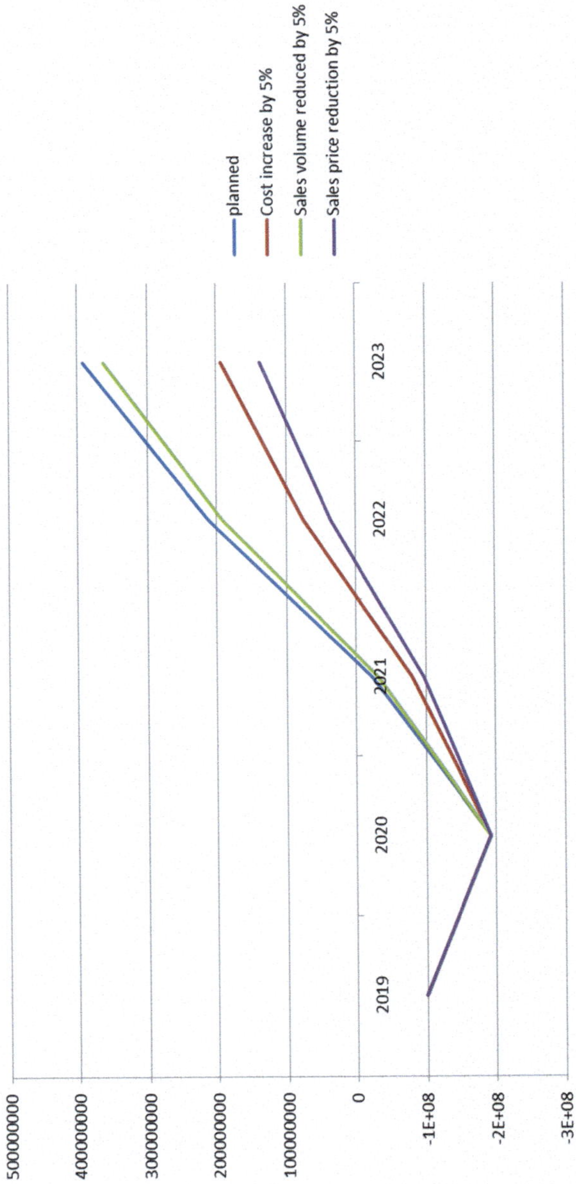

Chapter 6
How to Apply Cost Engineering Methods to Reduce Component Costs

6.1 Value Analysis and Value Engineering (VAVE) Guided by Standardization Questions

Value Analysis and Value Engineering were described systematically and in detail in Chap. 5. What does standardized questionnaire guided value analysis and value engineering mean here?

Most readers who have attended a VAVE cost reduction workshop will most likely have had this experience: In the VAVE seminar room (or production facility, see Fig. 6.1, for example), there were several subsystems (and disassembled parts) on the table, each with an identity label (subsystem name, part number, material, weight). The moderator asked the participants: "Do you have any ideas for reducing costs?" Usually there is silence at this moment. Sometimes someone suggested "dropping this sound-absorbing foam". The person who objected said, "Sound absorbers are installed by the NVH department, we don't need them for the interior." Another person said: "Reduce the wall thickness of the inner door panels from 2.5 mm to 2.0 mm", to which one participant immediately said: "The wall thickness of 2.0 mm cannot be produced with the current injection molding machine!"

Someone had an idea, and immediately another spoke out against it.

Two hours had passed and the VAVE workshop had not produced a single viable cost-cutting idea!

How can you make a VAVE workshop productive? The author thought about this question for a long time and came to the conclusion that VAVE workshops need a workshop methodology: The questions in the VAVE workshop need to be standardized.

X. Liu, *Cost Engineering Systems for Manufacturing Enterprises*,
https://doi.org/10.1007/978-981-96-5234-1_6

Fig. 6.1 Author led a VAVE workshop on site at Great Wall Motor's Tianjin Haval H6 production line

6.1.1 The Methodology of the VAVE Workshop

Firstly, Value Analysis & Value Engineering is a creative process that requires a fair and open environment and atmosphere. The following principles must be observed in VAVE workshops:

- There is no best design, only better design
- Do not ruin a good proposal with "killing phrases" such as the following
  - Already decided by the boss
  - Due to quality problems
  - that the styling has to be done this way
  - because a certain vehicle model also has it
  - etc.
- Do not answer questions in the style of a headline, but as quantitatively as possible
- Don't spend time arguing why it doesn't work (can't be improved). Instead, work with the Value Analysis and Value Engineering teams to figure out how it can work.
- The goal is to find a product design that provides a good customer experience, has the quality customers expect and is more cost-effective.

Secondly, precisely because VAVE workshops are a creative process, they need to be attended by experts and people who have a wealth of experience in product development, product manufacturing, product cost, product quality, product sales and, of course, new employees who may have ideas for cost reduction that no one else has thought of.

Finally, systematic standardization issues from the areas of product function, materials, manufacturing processes, tolerances and standards, surface treatment, make-or-buy must be clarified, which leads the VAVE team to consider all aspects and think about whether there is a better design and manufacturing solution, which corresponds to the "standardized questionnaire guided Value Analysis & Value Engineering (VAVE)" approach described by the author in Chapter 3.

6.1.2 Practical Examples of VAVE

1. Product design

(1) Shift bracket (see Fig. 6.2)

The value analysis revealed that the product specifications were too high: The actuating force of the gear knob was originally defined as 700N, which was far beyond the customer's normal range of use, and was reduced to 450N in consultation with the engineers in the chassis design department. At the same time, the design of the sheet metal part took into account not only its strength, but also its effect on the overall vehicle stiffness (NVH aspect) by designing reinforcement grooves into the shift bracket and creating a detailed plan for tooling and sheet metal part validation.

The end result was a cost reduction of 45 CNY per part and a weight reduction of 647 g. Based on the remaining life cycle of the 2.5-year vehicle model, the NPV of this cost reduction totaled CNY 10.12 million.

(2) Daytime running light module (see Fig. 6.3)

The LEDs were upgraded and optimized under the premise of ensuring luminous intensity and color rendering, and each daytime running light module was optimized, the cost was reduced by CNY 10.

(3) Intercooler fins (see Fig. 6.4)

The thickness of the fins was reduced from 0.12 mm to 0.08 mm, reducing the cost by CNY 3.38 per piece and the weight by 202 g.

(4) Damping mat on the rear body floor (see Fig. 6.5)

Through NVH simulation analysis, the yellow-marked area of the damping mat can be eliminated, reducing the cost by 2.32 CNY and the weight by 148 g.

The NPV calculation of the business case results in a total cost reduction of CNY 1.16 million.

2. Materials

(1) Exhaust pipe flange (see Fig. 6.6)

The material was changed from SUH409L to Q235 while ensuring performance (airtightness, etc.), resulting in a cost reduction of CNY 6.24 per piece (cost reduction for a single vehicle of CNY 37.44).

Fig. 6.2 Shift bracket for a passenger car. (Top: die-cast aluminum part; below: stamped and bent sheet metal part)

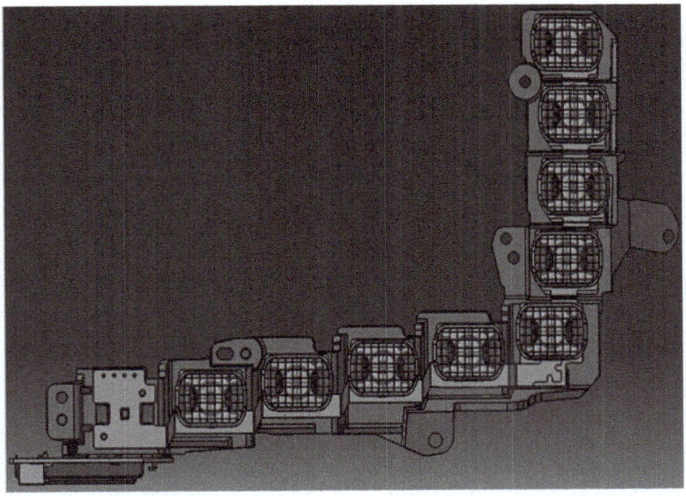

Fig. 6.3 Daytime running light module

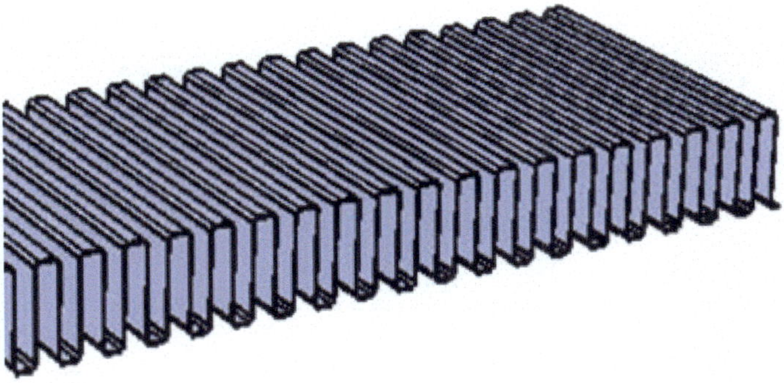

Fig. 6.4 Cooling fins of the intercooler

(2) Battery, starter and alternator wiring harness (see Fig. 6.7)

The conductor of the wiring harness was originally made of copper material; the copper conductor was replaced by an aluminum conductor. Due to the different specific electrical conductivity values of copper and aluminum, the cross-section of the aluminum conductor was increased by around 60% after the replacement in order to leave the resistance unchanged.

Since aluminum is much cheaper than copper, despite the increased volume of the aluminum conductor, the cost is still reduced by 2.10 CNY per vehicle, and the weight is reduced by 141 g.

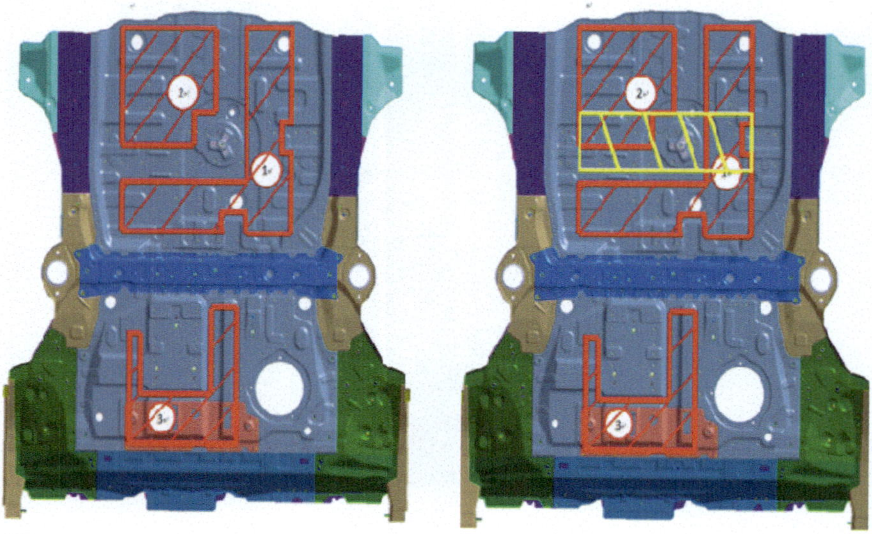

Fig. 6.5 Damping mat for the rear floor

Fig. 6.6 Exhaust pipe flange

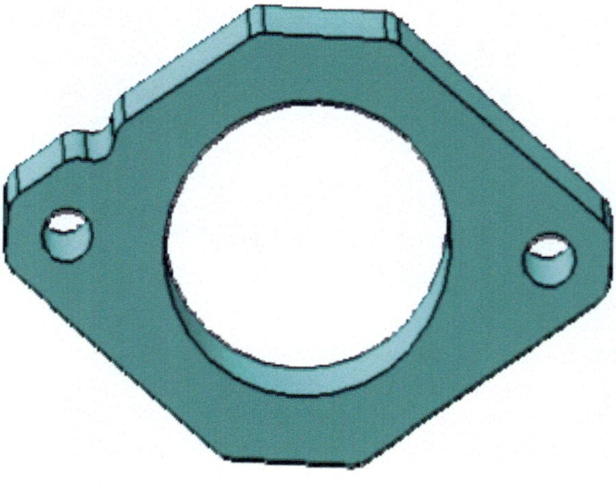

3. Production process

(1) Lower parts of the front and rear bumper (see Fig. 6.8)

Under the premise of ensuring the visual appearance and surface quality requirements of the components, the three spraying steps for primer, basecoat and topcoat were optimized into two spraying steps for primer and basecoat by optimizing the basecoat

Fig. 6.7 Battery, starter and generator wiring harness

Fig. 6.8 Lower parts of the front and rear bumper

color, which reduced the material cost and process cost of the spraying process per vehicle by 30 RMB.

(2) Dry hard turning instead of wet grinding (see Fig. 6.9)

Gear wheels for heavy trucks, dry hard turning instead of wet grinding, cost reduction of €0.35 per piece, business case calculation of NPV results in a total cost reduction of €300,000.

(3) Minimum quantity lubrication for machining large integral aircraft components (see Fig. 6.10)

Integral components of aircraft are usually manufactured in large, open machining centers (e.g. portal milling machines), which require large quantities of cooling and lubricating emulsions, consuming around 3000 L per hour, which, in addition to high costs, represents an enormous environmental burden.

Fig. 6.9 Dry hard turning instead of the wet grinding process

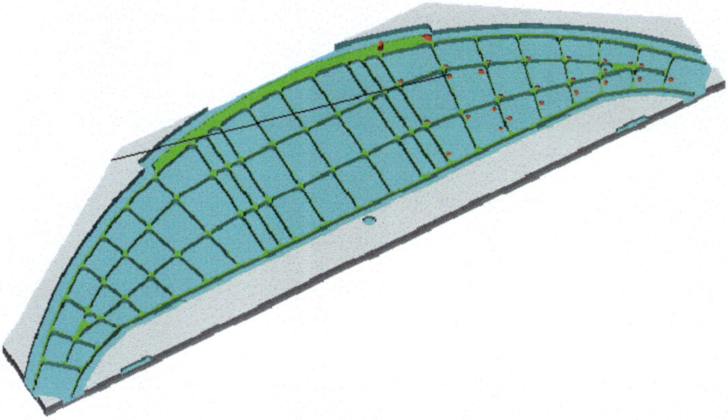

Fig. 6.10 Minimum quantity lubrication: machining of large integral aircraft components (forged aluminum fuselage and floor)

After almost three years of process development and component testing (e.g. compliance with manufacturing tolerances and surface quality, as well as avoidance of soft spots in forged aluminum alloys), the successful use of minimum quantity lubrication (with only 30 ml of cooling and lubricating fluid per hour, compared to 3000 L per hour previously) has both drastically reduced the environmental impact of handling large quantities of cooling and lubricating emulsion and lowered manufacturing costs, saving €24,000 per day.

One example, the minimum quantity lubrication process developed by colleagues and the author at the Daimler Group Research Center for the machining of large integral aircraft components, went into series production in 1996 at Airbus and was awarded the prize for new environmentally friendly processes by the city of Ulm.

6.2 Avoiding Over-Engineering

First a practical example:

The air conditioning control panel in a car (see Fig. 6.11): This is an air conditioning control panel from five different vehicle models, which fulfills the following customer functions:

- Switching the air conditioning on or off
- Setting and displaying the temperature of the air conditioning system
- Setting the airflow direction and airflow rate
- Switching the air recirculation on and off, defrosting the front and rear windows

  (1) The first model has 3 buttons, 9 switches (including LEDs to indicate the switching status) and 2 LCD displays
  (2) The second model has 11 push-button switches (including LEDs to indicate the switching status) and 1 LED display bar with symbols
  (3) The third model is equipped with 1 rotary knob and 14 push-button switches (some with LEDs to indicate the switching status).
  (4) The fourth model uses 2 rotary knobs and 10 switches (some with LEDs to indicate the switching status)
  (5) The fifth model uses only 3 knobs and 3 switches (with LEDs to indicate the switching status).

The price difference between these five models can be up to 200 CNY, apart from the influence of styling, which is a typical symptom of "over engineering". Over-engineering has the following typical symptoms:

- Complex realization of the product's customer functions.
- Product performance exceeds the customer's normal range of use
- Component service life is significantly longer than the product service life.

Product costs increase without any added value having been generated for customers, this is "over engineering".

It is imperative to design the product in such a way that the above-mentioned over-engineering symptoms are avoided during product specification and design.

Fig. 6.11 The climate control panel for the five different vehicle models

6.2.1 Product or Component Specification

The table of contents of a product or component specification is usually as follows.

1. Raguintroduction
2. requirements for vehicle integration
3. validation
4. quality and reliability
5. maintenance, servicing and repair
6. manufacturing processes
7. requirements for system suppliers
8. NVH
9. Passive safety requirements
10. Corrosion
11. Environment
12. Notes
13. appendices.

The first starting point for cost engineering is the product or component specification.

When reading the following sections of the product specification, please focus on Chap. 2 "Vehicle integration requirements" and Chap. 3 "Validation". Cost engineers must learn to read and understand the product or component specification, as this is the core of the technical optimization of product costs.

Some product specifications are very extensive (several hundred A4 pages), so here is an example of a relatively simple electronic vacuum pump (a subsystem) for passenger cars to illustrate the approach. The product specification of the electronic vacuum pump is as follows (extract):

1. *Introduction*

   This document specifies the requirements for the function, performance, reliability and validation of the electronic vacuum pump. The electronic vacuum pump is a functional subsystem that supports the brake booster function.

   ……

2. *Requirements for vehicle integration*

   The supplier is responsible for designing components with a minimum system weight and small dimensions in order to meet the customer's requirements and legal specifications.

   The electronic vacuum pump generates a permanent working vacuum in the brake system. The electronic vacuum pump, including the supplied contents, is designed by the supplier to optimize weight and costs in particular.

For the brake application, the fatigue strength of the electronic vacuum pump must be 1.2 million switching operations, one switching operation means the electric motor of the electronic vacuum pump switches on and off once.

The "solid state" relay is recommended for switching the electronic vacuum pump on and off

……

3. *Validation*

   This section defines the activities that must be carried out to validate the function and performance of the electronic vacuum pump.

   The following tests must be considered.

   a. *Durability test*
   b. *A service life of 1.2 million cycles is assumed for the 3.6 liter brake.*
   c. *Environmental tests*
   d. *Air to air temperature shock: 100 cycles between 120 °C and – 40 °C, 1 hour immersion; high pressure water spray: IPX9K protection class*
   e. *Salt spray*
   f. *Electronic vacuum pump: 144 hours according to DIN ISO 9227; electronic vacuum pump holder: 240 hours according to DIN ISO 9227*
   g. *Temperature fluctuations in accordance with DIN EN 60068, number of cycles 35; temperature changes between – 40°C and + 100 °C*
   h. *Mechanical tests*

Shock test: 10 shocks in each axis x, y and z in negative direction / 10 shocks in positive direction according to DIN EN 60068
Drop test: Must function properly when dropped from a height of 1 m onto concrete if there are no signs of external damage
Tension on the cable set: 90 N for 1 minute
Electrical test:
Overvoltage test: 17 V for 60 minutes at 25 °C
Jumper start test: 27 V, 20 s cycle time, 2 s on, 18 s off, 6 cycles
Insulation resistance: 500 V DC with motor running between plug and shaft
……

4. *Quality and reliability*

   This section is defined in the general specifications. If the current document is not listed in this section, the above specification applies.

5. *Maintenance, servicing and repair*

   This section is defined in the general specification.

6. *Manufacturing process*

   This section is essentially defined in the General Specification.

7. *Requirements for system suppliers*

   The electronic vacuum pump is defined by the chassis department and developed by the supplier. Certification and overall vehicle type approval is the responsibility of the Chassis and Vehicle Integration department. The supplier is responsible for all certifications and tests, reports, drawings, data sheets at subsystem and/or component level. The supplier is responsible for ensuring that the legal regulations applicable to certification/type approval are complied with.

   ......

8. *NVH*

   See General Specification

9. *Passive safety requirements*

   See General Specification

10. *Corrosion*

    The most important requirements for corrosion protection can be found in the "Specification book - Corrosion". In order to check the corrosion protection of components on the complete vehicle, the components must pass corrosion resistance tests.

    ......

11. *Environment*

    Environmental information such as resource conservation and recycling, materials and substances, limit values and evaporative emissions, toxicology and odors are listed in the instructions for use.

12. *Notes*

    Abbreviations and symbols

    List of materials

    General technical data

    EE specifications

    ......

Table 6.1 Over-engineering points from the product specification

Over-Engineering point	Cost optimization potential
Chapter 2 Requirements for vehicle integration: For the brake application, the fatigue strength of the electronic vacuum pump must be 1.2 million switching operations	Conversion to electronic vacuum pump with a service life of ≥ 250,000 switching operations, cost reduction of 300 CNY
Chapter 2 Requirements for vehicle integration: the "solid state" relay is recommended for switching the electronic vacuum pump on and off	A mechanical relay instead of the solid-state relay also guarantees a service life of ≥ 250,000 switching operations, cost reduction of 40 CNY
Chapter 3 Validation Tension on the cable set: 90 N for 1 min	Is there a possibility of such stress on the wiring harness during installation or during normal operation of the electronic vacuum pump? Can the diameter of the connector and the conductor be reduced?
……	……

13. *Appendices*

    ……

The following over-engineering points can be derived as examples from this product specification for electronic vacuum pumps (see Table 6.1).

Excessive technical designs are often the starting point for cost optimization. It is very helpful to read the product specification carefully and fully understand what is behind it.

6.2.2 Performance Requirements that Go Beyond the Customer's Normal Area of Application

Section 6.1.1 describes an example from VAVE practice: the gearshift holder (see Fig. 6.2), where it is mentioned that it was originally designed on the assumption that the customer would operate the gearshift knob with a manipulation force of 700 N (approx. 71.43 kg)! This raises the question: Does the driver (the customer) need such a force when operating the gear knob of an automatic transmission? Is this the customer's normal range of use?

No. In normal driving, the manipulation force should be less than 350 N. In other words, the designer has doubled the customer's normal operating range when developing this component. This is a classic case of over-engineering. The part was finally designed with the assumption of the manipulation force of less than 450 N so that both the cost was reduced by 45 CNY and the weight of the part was reduced by 647 g.

6.2.3 Component Lifetime Beyond the Product Lifetime

The electronic vacuum pump shown in Fig. 6.12 has an electric motor (component) that is designed to be started 1.2 million times, which meets the original requirements for product life (electronic vacuum pump).

If the e-motor is started every 1 km, the vehicle can travel 1.2 million km (in terms of the electronic vacuum pump alone), while the service life of a normal passenger car is 250,000–350,000 km. In other words, the service life requirements of this electronic vacuum pump far exceed the service life requirements of the vehicle. Another typical example of over-engineering.

Under the premise of guaranteeing the function/performance of the electronic vacuum pump and the NVH requirements of the vehicle, this service life requirement was reviewed again and reduced from the original 1.2 million e-motor starts to 450,000 (equivalent to 450,000 km in normal vehicle use) and validated accordingly. This resulted in a cost reduction of CNY 300 per unit and an NPV of CNY 22.5 million.

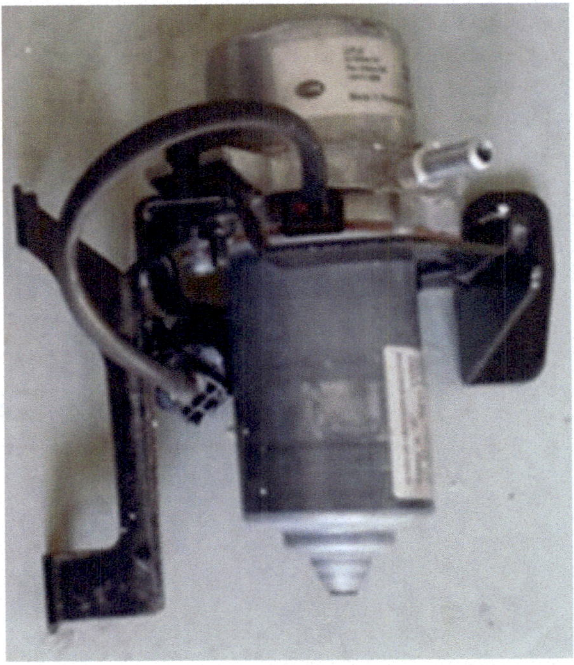

Fig. 6.12 An electronic vacuum pump

6.2.4 Practical Examples to Avoid Over-Engineering

1. Cross member of a front subframe (see Fig. 6.13)

The original design is as follows:

The body of the cross member is made of cold drawn seamless steel tube No. 35 with yield strength of 315 MPa, wall thickness of tubes 3 mm, the design safety factor is 3.0, which is typical over-engineering.

The optimized design is:

The body of the cross member is made of welded QStE380 steel pipe with a yield strength of 380 MPa, wall thickness of the pipes 2 mm, the design safety factor is 2.0, resulting in a cost reduction of 7.18 CNY per unit and a weight reduction of 1.12 kg.

2. galvanizing of the body (see Fig. 6.14)

To prevent corrosion, some sheet metal parts of the car body are coated with zinc on both sides. There are different coating thickness standards here, ranging from 40 g per square meter to 50 g or 70 g per square meter.

For most vehicle models with a zinc coating of 40 or 50 g per square meter, the vehicle will not rust over the life of the product under normal conditions. However, some vehicle models use 70 g per square meter of zinc coating.

If some body parts that do not need to be galvanized, e.g. interior parts, are also made from galvanized sheet metal, this can also lead to over-engineering. In this case, the solution to avoid over-engineering and to optimize costs is to omit the zinc layer from the interior parts (see Fig. 6.14 green parts) and to reduce the zinc layer

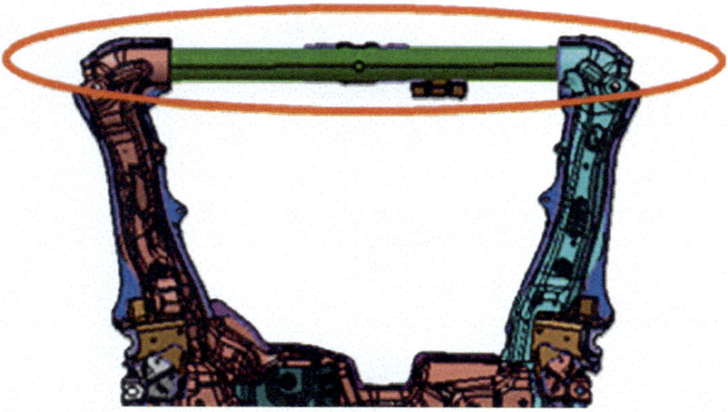

Fig. 6.13 Cross member of a front subframe

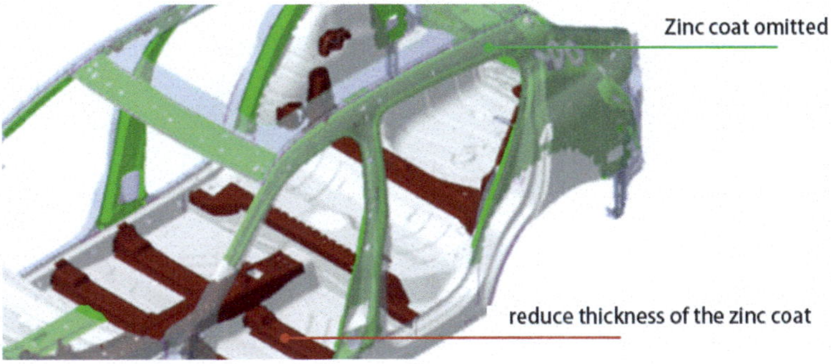

Fig. 6.14 Part of the body of a passenger car

thickness on the parts (see Fig. 6.14 brown parts) from 70 to 50 g per square meter. This reduces the costs per vehicle by around CNY 200.

6.3 Cost Engineering and Lightweight Construction

In order to reduce fuel consumption or electrical energy consumption, lightweight construction of the entire vehicle is a compelling trend. For example, to reduce the weight by 1 kg, a German automobile company is willing to increase the cost by 10 EUR (about 75 CNY), and a Japanese automobile company is willing to increase the cost by 5 EUR (about 37.50 CNY).

When it comes to lightweight construction, the first thing that comes to mind is aluminum or magnesium alloys, high-strength steels or expensive composite materials such as carbon fiber-reinforced plastics. The use of these materials increases material costs and complicates the manufacturing process (and therefore also manufacturing costs).

Given the current volatile sales and fierce competition in the automotive market, introducing lightweight construction on the basis of higher vehicle costs is no longer an option for most car manufacturers.

This is particularly true for new electric vehicles, where the weight has been reduced by replacing the combustion engine with an electric motor, but has been significantly increased by the installation of heavy battery packs (hundreds of kilograms) and heavier subsystems such as the high-voltage wiring harness, inverter and on-board charger. Lightweight construction and cost reduction are therefore also mandatory issues for electric vehicles.

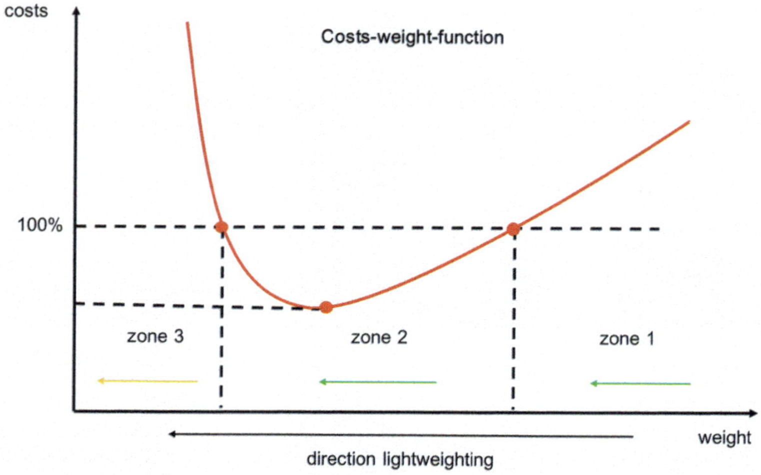

Fig. 6.15 Cost-weight function

6.3.1 Cost-Weight Function

How can weight be reduced and costs cut at the same time?

Let's start with an analysis of costs as a function of weight (see Fig. 6.15). In Fig. 6.15, the vertical axis is the cost, the horizontal axis is the weight, and the horizontal axis can be divided into three zones. In the first zone, the costs decrease with the weight, and in this zone the weight is still relatively high. In the second zone, the weight is reduced further and the costs are lowered to a point where the costs are at their lowest, which is the optimum for lightweight construction. The weight is then reduced further, but the costs increase again and the horizontal axis moves from the second zone to the third. In the third zone, the weight reduction continues, but the costs also increase sharply.

6.3.2 Practical Examples of Cost and Weight Reduction

Below you will find some practical examples of how weight and costs can be reduced at the same time.

1. Optimization of the component geometry

In the first area, the geometry of the components is optimized in order to reduce weight and costs while using the same material. Figure 6.16 shows an example where the material is PP+EPDM+TD20. In the curve of the functional relationship

Fig. 6.16 Lower left bumper

between weight and cost, the design of the component is still in the first zone. The component geometry can be optimized geometrically by using the CAE tool (topology optimization). The optimized geometry is, of course, the one that must be analyzed for manufacturability (e.g. here with the help of process simulation software for injection moulding).

In this example, the wall thickness of the component was reduced from 3.0 to 2.50 mm, the manufacturability was ensured by a process simulation, so that a cost reduction of 2.70 CNY was realized, at the same time the weight of the component was reduced by 114 g.

2. Optimization of the design safety factor

The focus here is on performing a CAE strength analysis based on the actual load spectrum on the component and comparing the relevant parameters of the materials used. As the material parameters used in the CAE analysis or finite element calculation do not match the actual material 100%, a so-called design safety factor is used when designing the component geometry. In other words, the actual load spectrum and the design safety factor are the key parameters for lightweight construction.

When the lightweight construction enters the second area (see Fig. 6.15), the weight is further reduced. The first point that needs to be analyzed in detail is the design safety factor.

If it is assumed that the material property curve used in the finite element calculation deviates ± 15% from the actual material properties, and the finite element calculation usually has ± 10% calculation deviation, the optimal design safety factor for the structure is 1.25.

In order to further reduce the weight and costs, the design safety factor was reduced from 3.0 to 2.0 in the practical example (see Fig. 6.13) (there is still a margin to the ideal design safety factor of 1.25), which led to a reduction in weight and costs.

According to experienced chassis experts, the optimization of the design safety factor can start at 1.5 and be optimized to 1.3 in two or three iterations.

3. Optimization of materials

When we talk about lightweight construction, the first thing that comes to mind is a change of material. If aluminum is used instead of copper, aluminum has a higher specific electrical resistance than copper and the cross-section of the aluminum

Fig. 6.17 Battery holder

conductors must be increased (by about 60%), but the density of aluminum is 2.7 g/cm^3, while that of copper is much higher (8.9 g/cm^3), and the cost of aluminum is much lower than that of copper. The lower weight therefore also reduces the costs (see Fig. 6.7).

Another example: The battery holder, which normally consists of several sheet metal parts, was replaced by glass fiber reinforced plastic (PP+LGF30), which reduced the weight by about 500 g and the cost by about 20 CNY (see Fig. 6.17).

With further weight reduction, the third zone is reached (see Fig. 6.15), which is characterized by the use of fibre-reinforced materials and in which the costs increase rapidly.

Near the transition between the second and third zones is a practical example of a plastic rear door (see Fig. 6.18), which consists of inner and outer panel and frames made of plastic, with the hinges and some reinforcements still made of metal. With this design, it is possible to reduce the weight by about 7 kg, with a cost increase of 0–150 CNY (compared to a pure sheet metal rear door). By optimizing the painting process and reducing the scrap rate, a weight reduction of 7 kg can be achieved with the plastic rear door at the same cost.

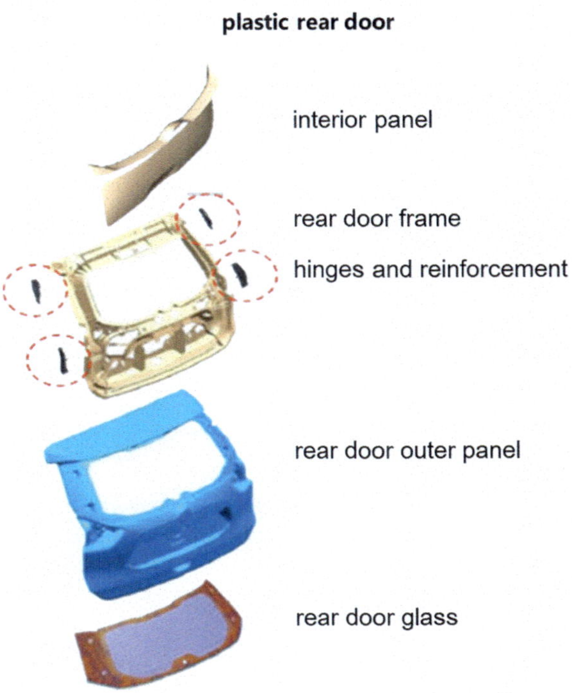

Fig. 6.18 Plastic rear door

Chapter 7
The Target Cost Process of a Product

The target cost process of a product is the central process in a manufacturing company. To illustrate this, here is a true story:

A former wholly owned subsidiary of the Daimler Group, Dornier Luftfahrt, used a lot of new technology at the time to develop the Do328 (see Fig. 7.1), a civilian aircraft with twin turbopropellers and later twin turbines, to develop a "Mercedes in the air" with a long list of product innovations, e.g. the glass cockpit, head-up display, etc. required significant research and development funds and high material costs.

When these costs and extremely optimistic market forecasts (projected sales volume and projected sales price) are entered into the business case analysis, the calculation of the project's NPV shows that it will take 300 Do328s sold to reach the break-even point in the NPV curve, that means, it will take 300 aircraft sold at the projected sales price before the one-time R&D expenses and investments are recouped.

Unfortunately, this was not to be:

- demand in the aviation market for a 30-seat regional aircraft like the Do328 was much lower than forecast;
- the forecast sales price could not be realized either, the actual price was significantly lower than forecast;
- the enormous expenditure on research and development had to be allocated to a much lower number of sales.

Ultimately, this led to a loss of DM 110,000 per Do328 sold. Dornier Luftfahrt was in turn forced to continue producing and selling this civil aircraft in order to maintain the production site and jobs in Germany. The Do328 was produced in a total quantity of 217, and in the end Dornier Luftfahrt lost so much money that the company changed hands several times and finally went bankrupt.

This story of Dornier Luftfahrt shows us that:

X. Liu, *Cost Engineering Systems for Manufacturing Enterprises*,
https://doi.org/10.1007/978-981-96-5234-1_7

Fig. 7.1 The Do328jet aircraft

- An accurate, realistic market analysis and product definition is essential, in particular a realistic sales and price forecast as well as the functions/services desired by customers
- An accurate business case calculation (NPV) of the project and the NPV sensitivity analysis in relation to the achievement of the target costs, the forecasted sales volume, the forecasted sales price etc. is required.
- Product development must pay attention to costs. Gone are the days when R&D engineers developed the product first and then let the cost engineers calculate the costs. The "art of engineering" is not only to implement the functions/services of the product, but also to realize them at the defined target costs and target quality.

The blue bars in Fig. 7.2 indicate the tasks of cost engineering in the product process, also known as the "target cost process". The target cost process is divided into four phases.

- Determination of the target costs
- Design at the target costs (design to cost)
- Control of engineering changes
- VAVE in series production.

They are important components of the product process. A product is not primarily technology-oriented, but first and foremost market- and customer-oriented, so that the company has a sustainable future.

(1) Target cost definition:

After entering the forecast sales volume, the forecast sales price and the one-time expenses (R&D costs, tooling costs, investment in the production line) as well as the target profit margin, Cost Engineering determines the target costs for the product. Based on an initial bill of materials for the product, the target costs are then broken down to the subsystems or components.

(2) Design to cost:

Before the product data is frozen, design-to-cost methods are applied (e.g. morphological box, function-cost matrix). The development departments design subsystems

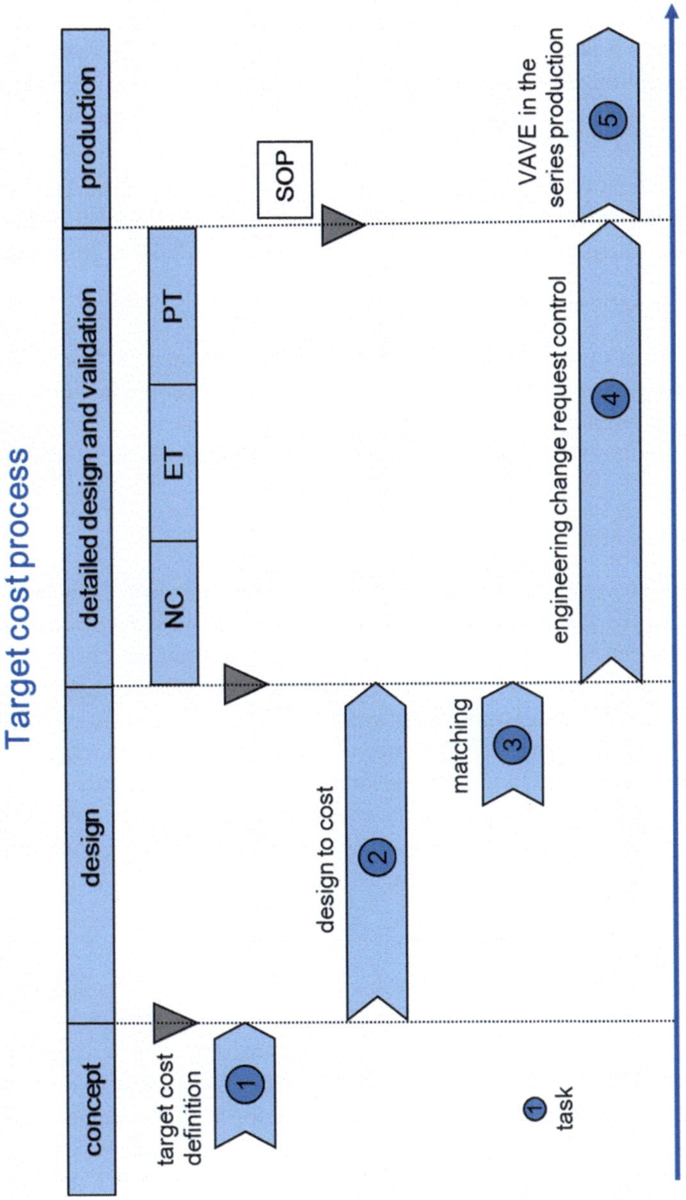

Fig. 7.2 Product target cost process

or components at the respective allocated target costs, whereby the likely achievable costs must be assessed on the basis of the technical design. This phase is also the best time to optimize product costs.

(3) Control of engineering changes:

During the validation phase, changes are made to the designs of certain subsystems or components. The key is to evaluate the cost variance for each cost-impacting engineering change and try to find cost-neutral solutions for the engineering changes.

(4) VAVE in series production:

Continuous cost optimization and efficiency improvement is a "norm" for a good manufacturing company, where daily cost optimization becomes a daily business for the manufacturer, not just a temporary cost reduction project.

Here the author shares two of his experiences at Mercedes-Benz and Great Wall Motor:

(1) Mercedes-Benz Germany

With every new vehicle model development come engineering changes, thousands of them every year. In one case, the R&D department proposed a design change that the cost engineering department estimated would increase the cost by €0.27 and the tooling change cost by €110,000. The project manager then asked in a project meeting: "Is there a change option that would not increase the costs by €0.27?" "Do the tools have to be changed?" The three departments—R&D, Purchasing and Cost Engineering—then spent four weeks debating whether a cost increase of €0.27 on the component and €110,000 on the tooling was justified. On the face of it, Mercedes-Benz may not be short of cash, but such a cautious approach is commonplace, and we're "only" talking about €0.27 here.

(2) Great Wall Motor China

One day, a chassis development engineer suggested a engineering change to the rims because the natural vibration frequencies of the tires and rims were too close to each other and tended to resonate. The change consisted of increasing the mass of the rims and changing the modal frequency so that the natural frequencies of the two components were further apart. In the discussion with this engineer, the author asked the question: "Changing the modal frequency, there is one way to increase the mass, have you investigated the second way to decrease the mass?" In response, the engineer said, "Ah, why didn't I think of that?" Ultimately, the engineering change was to reduce the mass of the rims while ensuring strength and comfort in a way that avoided resonance between the tires and rims while reducing the cost and weight of the rims compared to the pre-design change.

7.1 The First Customer

Some leading companies have a "first customer" department in R&D, which is similar to a "blue army unit" and is dedicated to looking at the product definition and product development content through the eyes of the customer and providing improvement advice to the development department.

An example of the First Customer's Eye View for a new vehicle model is shown in Table 7.1:

In the product definition phase, the first step is to carry out a "customer requirements analysis". The surveyed customers and the "First Customer" will request improvements to the product's features/services (see Fig. 7.3), and in response to these customer requirements, measures will be defined for each of the new vehicle models to be developed.

In addition to the input of the "first customer", it is even more important to listen to the opinions of real customers. Most car manufacturers carry out a so-called "Car Clinic Study" before defining the product of a new vehicle model. In the car clinic, the opinions of real customers must take precedence over those of the company's CEO or board members.

Table 7.1 Eyes from the first customer

Category	Attention from the first customer
Space available	Front seats, knee room, space from head to floor; armrests and storage box, glove compartment and ease of use
	Seat comfort, adjustability, seat heating
	Rear seat comfort, knee room
	Trunk size, easy loading and unloading of luggage
Passive and active safety	C-NCAP crash results, front airbags, side airbags, window bag, knee airbags
	Brake assist, emergency braking radar (millimeter wave radar), reversing assist
Man–machine interface	Easy operation and visibility of
	Navigation, instrument cluster, interior and exterior mirrors, etc.
Air conditioning, heating	Climate control and heating Comfort, reaction speed
Comfort	Ride comfort, NVH damping
Handling	Straight-line stability, steering, acceleration, braking
Odors	Odors, VOC in the cabin

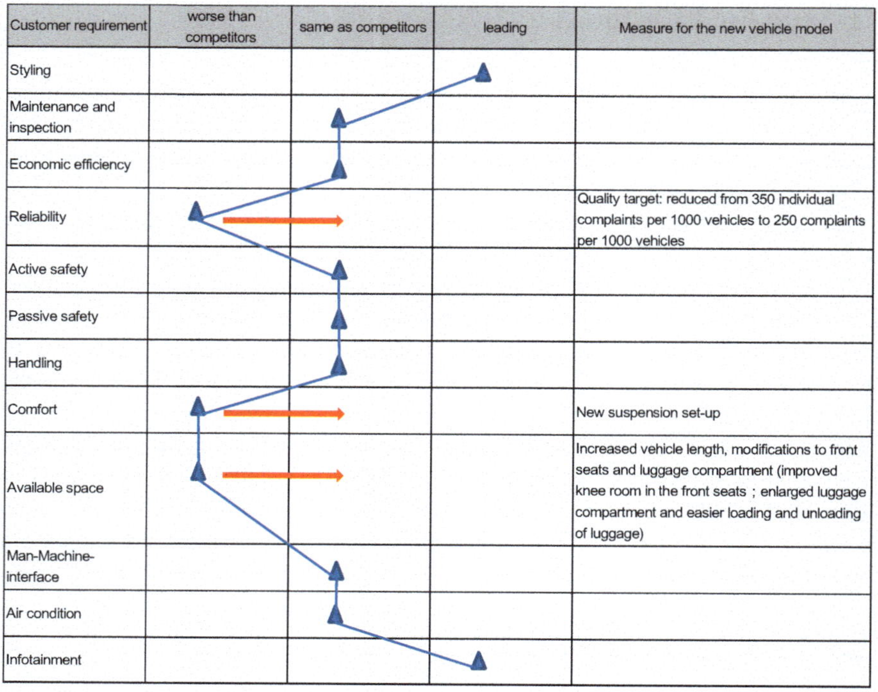

Fig. 7.3 Analysis of customer needs and areas for improvement

7.2 Determining Target Product Costs

Determining target product costs is one of the core tasks of cost engineering and is an important deliverable that should be addressed at the end of the product concept planning phase and before the product project is launched.

7.2.1 Determining the Target Costs

The information in Table 7.2 is required to determine the target costs.

Other requirements:

- Target return 8%
- Capital interest rate of 7.5%
- One-off expenses (R&D, tools, investments in the production line)
- Provision for design changes (10%)

The target costs of the vehicle model are thus calculated according to the total vehicle cost model (Table 7.3).

Table 7.2 Inputs for determining the target costs of a product

Department	Inputs			
Product definition	Target profit margin on the product			
Sales	Forecast sales volume and forecast sales price as well as their changes over the planned product life cycle			
Product definition	Overall vehicle concept, option list			
Quality, customer service	Repair costs during the warranty period			
Cost engineering	Determination of target product costs, provision for engineering changes			
Determination of target costs using the example of an SUV				
Project code	XXX			
Sales volume	100.000 units/year			
Market launch	September 2021			
Sales forecast				
Time	2021	2022	2023	
Sales volume	30.000	90.000	75.000	
Sales price				
Car line/Price	Basis (CNY)	Middle (CNY)	High (CNY)	Pro (CNY)
1.5T + 6MT	100.000			
1.5T + 6DCT		120.000	140.000	
2.0T + 6DCT				160.000

Table 7.3 Target costs of the vehicle model

Car line	1.5T + 6MT Basis	1.5T + 6DCT Middle	1.5T + 6DCT High	2.0T + 6DCT Pro
Target costs (CNY)	70.542	75.672	80.788	85.918
Provision engineering changes (10%) (CNY)	7.054	7.567	8.079	8.592
Final target cost (CNY)	63.488	68.105	72.709	77.326

The 1.5T + 6DCT High equipment line is the main vehicle model and serves as the basis for breaking down the target costs of the overall vehicle into the various subsystems. Once the target costs for the entire vehicle have been determined, a business case calculation (NPV) can be carried out (see Fig. 7.4).

Fig. 7.4 Business case calculation (NPV) for an SUV

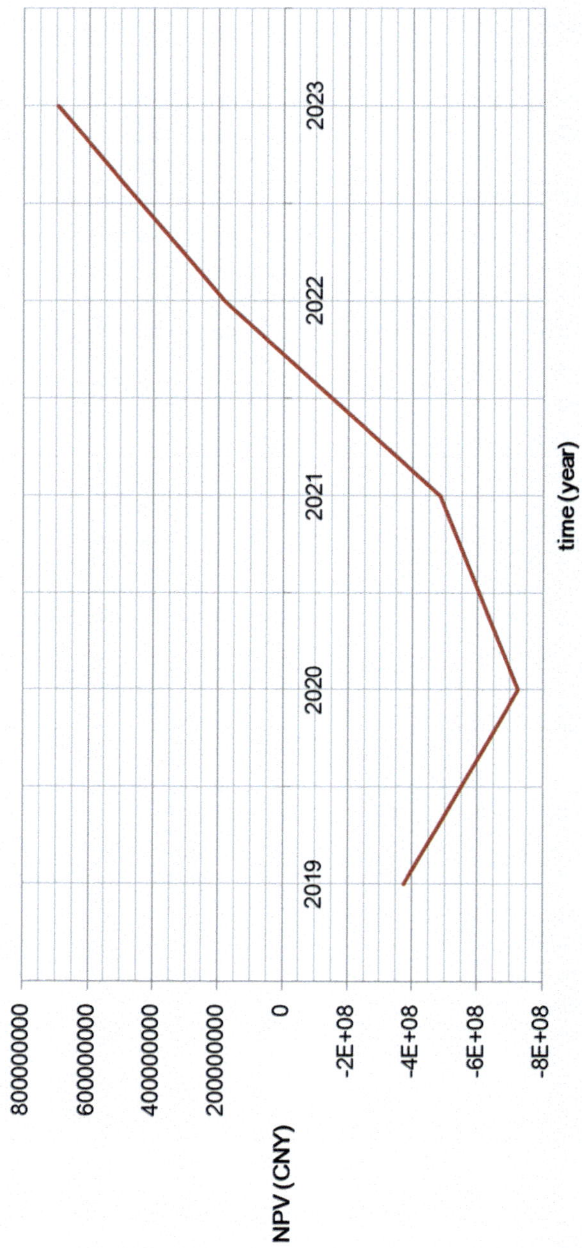

7.2.2 Subdivision and Adjustment of Target Costs

In the case of the SUV described above, in order to break down the target costs of the vehicle into the various subsystems and components, it is first necessary to create an "initial bill of materials" for the entire vehicle.

When new option is added to a new vehicle model, the initial BOM and the associated new subsystems or components must be added due to the option changes, resulting in a complete initial BOM.

Assuming that VC_t is the final target cost of the vehicle, VC_r is the cost of the reference vehicle of the series production, and C_r is the cost of a subsystem of the reference vehicle, the target cost of the corresponding subsystem C_t for the new vehicle model is linearly decomposed as

$$C_t = \frac{VC_t}{VC_r} C_r \tag{7.1}$$

Accordingly, the standard parts are not decomposed linearly and the original costs are retained. This results in the target costs for each subsystem of the new vehicle model (Table 7.4).

Table 7.4 Target costs for each subsystem of the new vehicle model

Subsystem	Source of supply	Quantity	Target costs (CNY)
Automatic transmission	External procurement	1	8414.83
AS Engine	Self-build	1	11656.52
Engine mount left	External procurement	1	32.68
Engine mount bracket left	External procurement	1	43.57
Connecting piece	External procurement	1	58.59
Engine mount right	External procurement	1	133.71
Rear engine mount bracket	External procurement	1	5.78
Rear engine mount	External procurement	1	44.68
Dual mass flywheel	External procurement	1	526.59
Fuel tank	External procurement	1	244.35
Protective bracket	External procurement	1	8.17
Carbon canister	External procurement	1	15.51
Fuel tank cap	External procurement	1	7.96
Fuel filler neck	External procurement	1	67.76
Protective cover	External procurement	1	2.82
Worm gear rim	External procurement	1	1.25
Four-hole hose clamp	External procurement	3	1.67
Outer oil inlet hose	External procurement	1	9.59
……	……	……	……

Once the target costs for each subsystem have been broken down, the next important task for the development and cost engineering department is to design to cost. The methods for this were described in detail in Chap. 5, e.g. morphological box and function-cost matrix.

Based on the technical designs for the subsystems, Cost Engineering creates a cost calculation (forecast) for each subsystem in order to evaluate the technical designs of the subsystems and compare them with the previously defined target costs of the subsystems. If the target costs will be achieved, the corresponding technical designs of the subsystems are approved for release by three parties (project manager, head of the design department, head of cost engineering) (see Fig. 7.5).

Since the cost of each system is generated by various requirements from the design, styling, sales, etc. departments, the total amount of cost reduction for the system in the design process must be shared by these departments, and the following is a practical example (Table 7.5).

If some subsystems go through multiple iterations (e.g. more than 3 iterations) of design optimization using design-to-cost methods and still cannot meet the target cost, a matching process is required.

In the matching process, the cost engineer performs an analysis of the functional/performance options of a subsystem to provide the project team with a clear cost breakdown of the options (see Figs. 7.6 and 7.7); the sales department is responsible

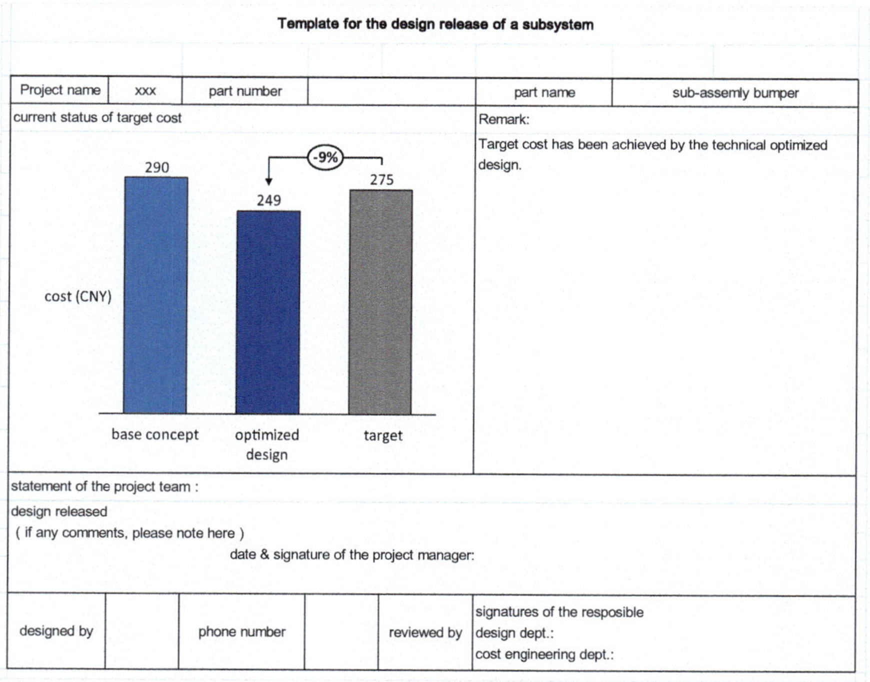

Fig. 7.5 Form for approving the technical design

Table 7.5 Common cost reduction targets of departments

System	Cost reduction target (CNY)	Departments involved
Body in white	800	Body shell, safety, styling
Interior and exterior	1.500	Interior and exterior, styling, NVH
EE	2.000	EE, sales
Powertrain	500	Powertrain, handling
Chassis	300	Chassis, handling

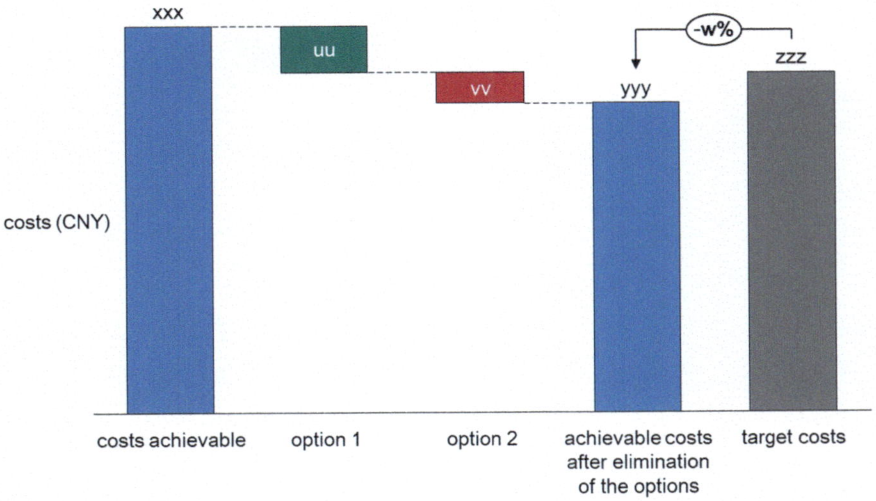

Fig. 7.6 Relationship between the achievable costs of the subsystem and the target costs (example)

for listing the customer's functional requirements and categorizing these functions as "important", "average" or "not important" for the customer. Finally, the project team decides which options can be omitted and which must be retained (see Fig. 7.7).

7.2.3 The Target Cost Achievement Path

How can the target costs for a product be achieved? The optimization potential of product costs is around 85% at the beginning of product development (see blue curve in Fig. 7.8) and only around 15% in the later phases of product validation.

As shown in Fig. 7.8, in the conventional product cost target trajectory (red curve), the starting point for the target costs is usually the actual costs of a series product or a reference product, or only a vague target cost range is defined. If the product project has already reached the validation phase, it turns out that the predicted product costs are too high when calculating the profitability, cost-cutting measures must be taken

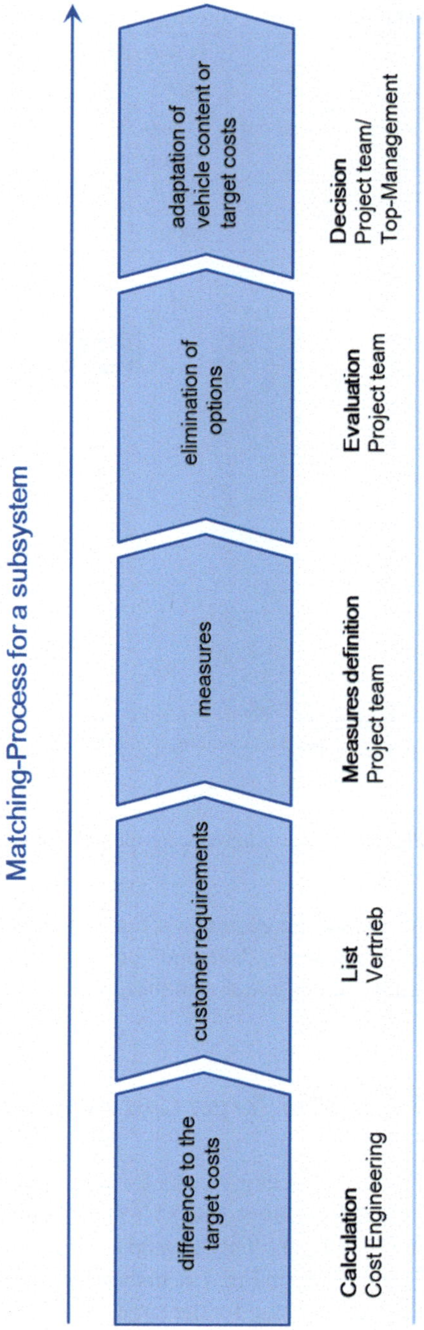

Fig. 7.7 Matching process of a subsystem

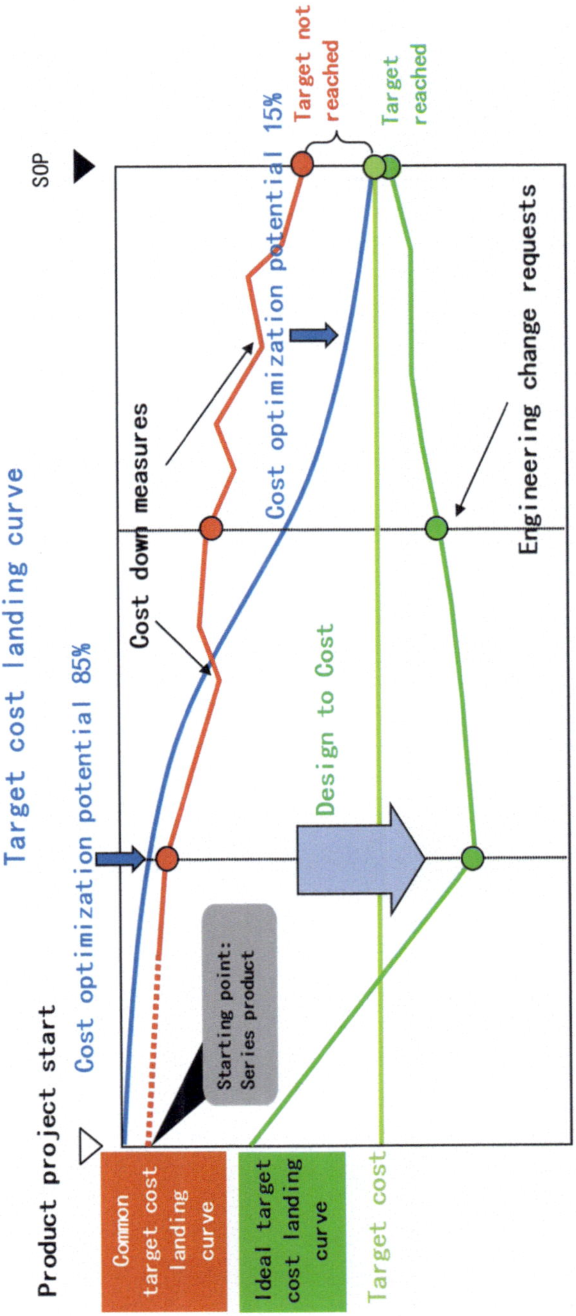

Fig. 7.8 The target cost achievement path of a product

quickly. Unfortunately, they are less successful because most of the designs have already been completed and considerable one-time costs would be incurred if these designs had to be changed at this stage. In the end, product profitability is much lower than expected.

To truly implement the target cost process (green curve), a clear product target cost must be set in line with the business case calculation (NPV), with a contingency for design changes (5–10% of the product target cost). The product and associated subsystems are designed to achieve the target cost and ultimately the expected NPV.

Chapter 8
Optimization of Manufacturing Costs

Figure 8.1 shows the cost components of a car bumper, indicating that manufacturing costs account for approximately 38%. Of course, different manufacturing processes and combinations of processes, different production facilities, different labor hours, etc. influence this percentage. In addition to the optimization of material costs, the optimization of manufacturing costs is also a "must" for manufacturing companies.

The main components of manufacturing costs are as follows (Table 8.1).

From a manufactory to mass production on an assembly line, the aim is to reduce unit costs and improve production efficiency and product quality. Over the decades, manufacturing companies have developed and created various corporate management structures, production systems and processes to reduce manufacturing costs and increase efficiency.

8.1 Toyota Production System (TPS)

The core elements of the Toyota Production System are:

- Elimination of all waste
- Continuous improvement.

Waste is anything that does not advance the process and anything that does not add value. Typical wastes in a production system are:

- Poor layout of the production line
- Low utilization of production facilities or unused production equipment
- Production materials (or components) are not optimally utilized, superfluous packaging
- Inconsistent quality requirements for production materials (or components) or lack of quantifiable quality standards, high return rates or reworking

X. Liu, *Cost Engineering Systems for Manufacturing Enterprises*,
https://doi.org/10.1007/978-981-96-5234-1_8

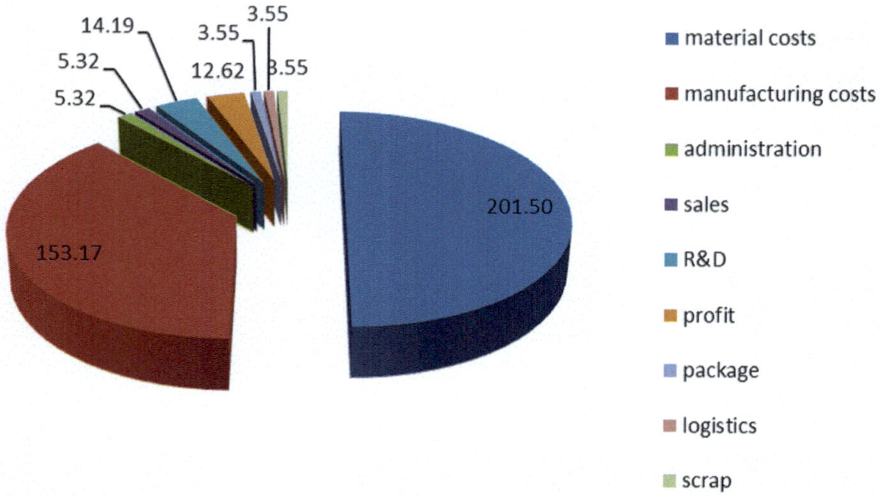

Fig. 8.1 Cost break-down of a front bumper for a passenger car

Table 8.1 Main components of manufacturing costs

Category	Cost type
Direct costs	Allocation of production facilities
	Labor costs
	Costs of energy, water, gas, etc.
Indirect costs	Maintenance costs
	Quality assurance costs
	Management costs of the production line

- Excessive inventory (waste of capital), overproduction (production exceeds customer demand)
- Inefficient transportation of personnel or excessively long travel distances
- Uneven distribution of employees at workstations, some are busy, others are idle
- Inferior end products
- etc.

According to the author, "there is no such thing as the best product design, there will be always a better product design". In the same way, in a production system "there is no most efficient production line, only a more efficient production line".

Continuous improvement is not about "developing a culture of constant problem solving, but about finding the cause of the problem and doing it right once so that the same problem does not occur again tomorrow"; it is also about every employee learning and improving their skills in order to achieve continuous improvement.

8.2 Mercedes-Benz Production System (MPS)

The author has personally experienced Mercedes-Benz over many years on several production lines (including car, van and truck production lines as well as engine, transmission, chassis and other subsystem and component production lines) and has also been involved for many years in the development of new production processes at the Group Research Center.

The Mercedes-Benz production line is known for its high quality requirements and automation and combines the ideas of waste elimination and continuous improvement of the Toyota Manufacturing System into a self-contained production system: the Mercedes-Benz Production System (MPS).

8.2.1 Value Stream Mapping

Value stream mapping, also known as material and information flow analysis, is a method to investigate material and information flows across the entire value chain in a uniform and cross-process manner, identifying cases of doubt and finding optimization points.

As shown in Fig. 8.2, the value streams are divided into:

- Material flow
- Information flow

The material flow refers to the entire process from raw materials and semi-finished products through manufacturing and assembly in the production line to the finished products. The information flow, on the other hand, refers to the process that ranges from market forecasts and customer orders to the creation of production plans and information about raw materials and finished products.

Value stream mapping consists of two steps. In the first step, the current status of the value stream is recorded:

- analyze the customer requirements
- describe the affected workstations and work content
- identify the point at which the process is triggered
- record the process step sequence
- present the process data, transfer points, information
- show the internal material flow, information flow
- calculate the runtime and value creation time.

In the second step, the "ideal" value stream is described:

- Consideration of customer demand and production quota
- Reduction and optimization of transfer points
- Optimization of production lines and production facilities

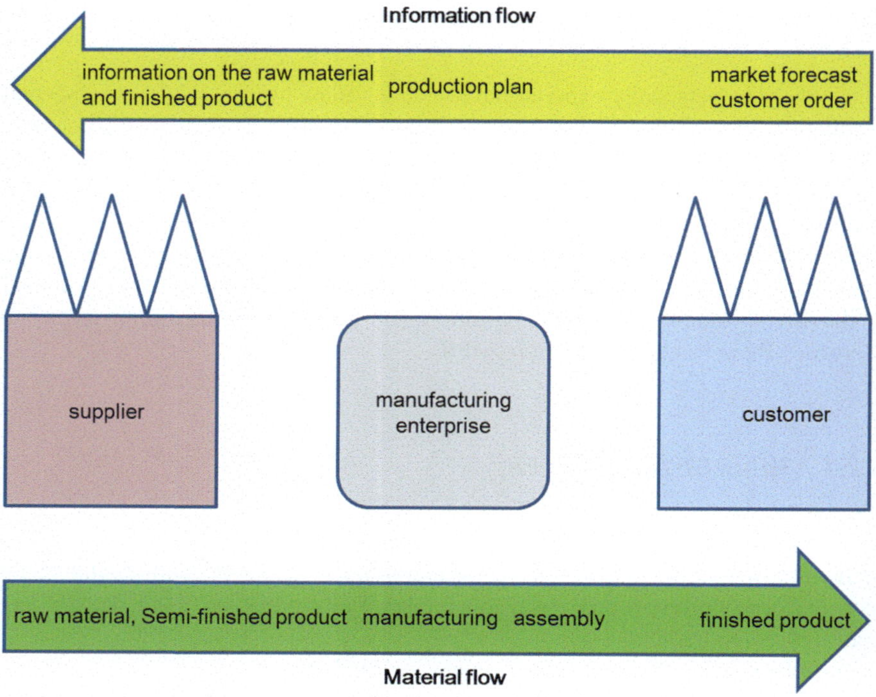

Fig. 8.2 Value stream mapping

- "Smoothing" of production processes
- Creation of a "pull" system
- Consideration of market fluctuations and production flexibility
- List of suggested improvements.

To illustrate the value stream analysis method, an example of a component assembly line is shown below. First, the current status of the value stream in this component assembly line is analyzed (see Fig. 8.3).

(1) Arrangement of the assembly line
(2) Distance between the workstations
(3) Detailed description of the individual processes and analysis of the working time
(4) Calculation of the assembly line cycle (dwell time at the longest station)

The results of the working time analysis are shown in Fig. 8.4.

As shown in Fig. 8.4, workstation 5 is the step sequence that determines the cycle of the assembly line with a time of 36 s, while stations 2 and 4 have to wait 10 s each and station 3 even 20 s, which means greater waste.

The next step is to identify waste reduction and optimization points to describe the "ideal" value stream (see Fig. 8.5).

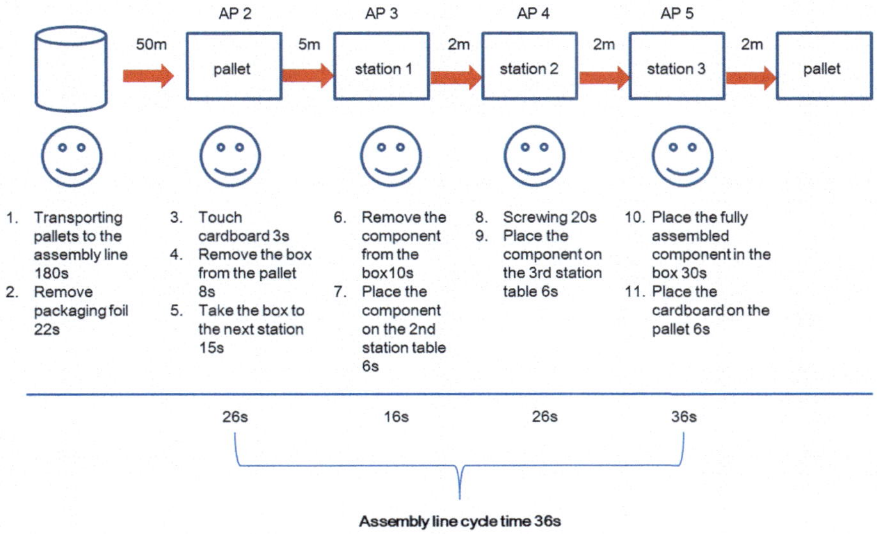

Fig. 8.3 Current status of the value stream of a component assembly line

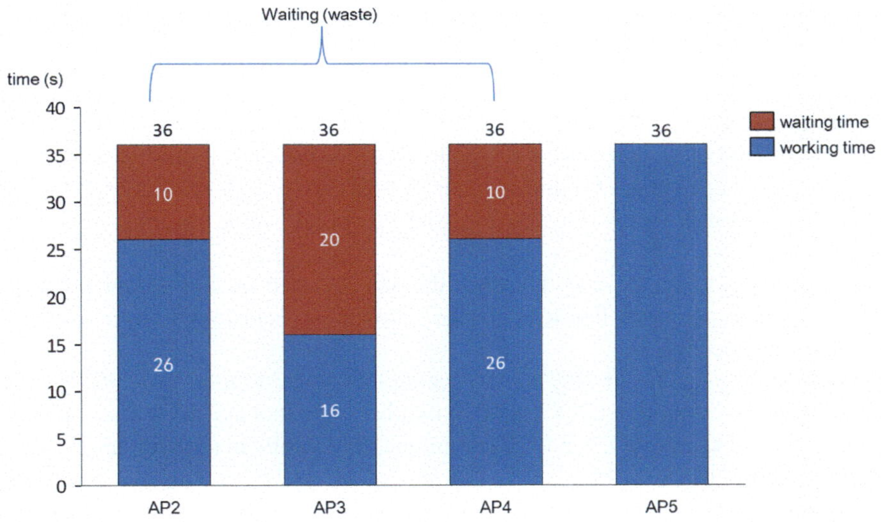

Fig. 8.4 Analysis of the current status of working times

- Changing the production line layout from "one straight line" to "U-shaped"
- Shortening the walking distances
- Optimization of processes 5, 7, 9 and 11
- Elimination of the pallet station workstation (AP2).

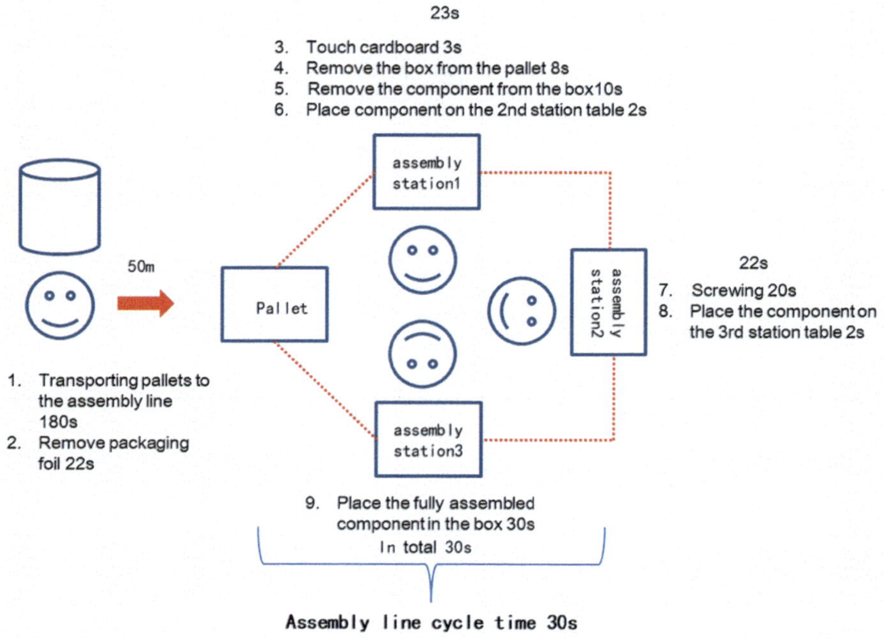

Fig. 8.5 The "ideal" value stream for a component assembly line

The assembly time was reduced from 36 to 30 s and the waiting time from a maximum of 20 s to a maximum of 8 s. The result of the analysis and optimization is the "ideal" value stream for this component assembly line shown in Fig. 8.5.

A comparison of the working times under the "ideal" value stream is shown in Fig. 8.6.

In summary, it can be said that the value stream analysis and optimization of this component assembly line leads to the following conclusions from a cost and efficiency perspective:

Productivity increase of 16.7% by reducing the assembly time from 36 to 30 s, assuming an investment of CNY 1 million in this component assembly line and thus the machine allocation of CNY 0.95 per minute, this results in a saving in machine costs per unit of CNY 0.095.

Elimination of one workstation (pallet station AP2), if the labor costs are 0.50 CNY per minute, the labor costs per unit are reduced by 0.25 CNY.

In total, the value stream analysis and optimization leads to a reduction in manufacturing costs per unit of $0.095 + 0.25 = 0.345$ CNY.

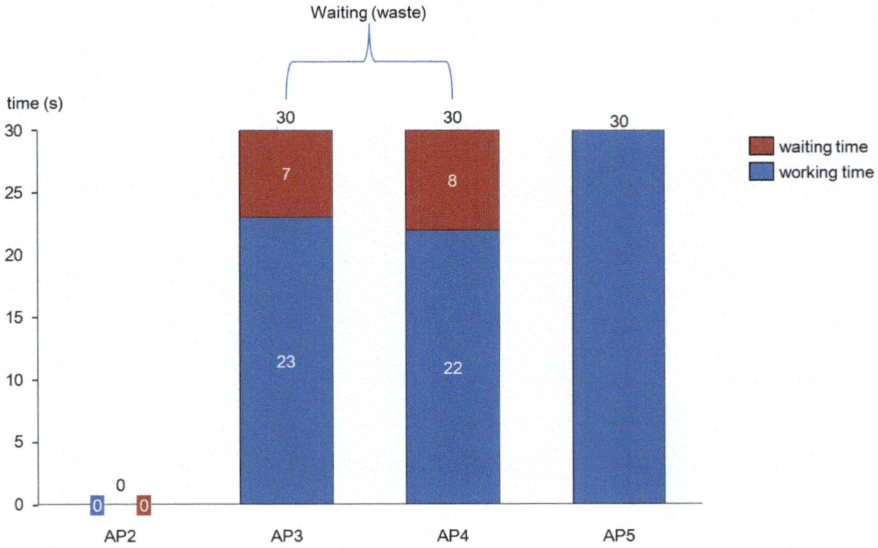

Fig. 8.6 Results after optimization of the operating time

8.2.2 *Method of Data Collection at the Workplace*

The prerequisites for the workplace data collection method are:

- Constantly repeating work processes with process cycles ≥ 1 min
- Relatively uniform repetition of work processes
- The process status can be better visualized and separated
- The observer is not the operator of the workstation.

The method of data collection at the workplace is to describe a process in terms of operations (e.g. screwing, removing screws, removing screwdrivers, walking, waiting ……), and the measured time interval is usually recorded in 10 s, and a vertical line is drawn.

During the analysis and optimization process, potential for improvement is sought by identifying the operations that add value and those that do not (Table 8.2).

Figure 8.7, which shows the results of the data collected, shows that there is an overall potential for improvement of up to 23% at this workplace

8.2.3 *7V (7 Types of Waste to Avoid)*

In general, work in manufacturing companies can be divided into three categories (see Fig. 8.8):

1. work that adds value (customer functionality, customer satisfaction)

Table 8.2 Method of data collection at the workplace

Cycle no	Assembly	Handling	Walking	Wait	Cleaning	Others	Total
1	IIIII IIIII	IIIII III	IIII	II	I		25
2	IIIII III	IIIII IIII	IIII	III	I		25
3	IIIII IIIII I	IIIII III	II	I	I	II	25
4	IIIII IIII	IIIII II	III	III	II	I	25
5	IIIII IIIII	IIIII III	IIIII	II			25
Quantity	48	40	18	11	5	3	125
Percentage	38.4	32	14.4	8.8	4.0	2.4	100

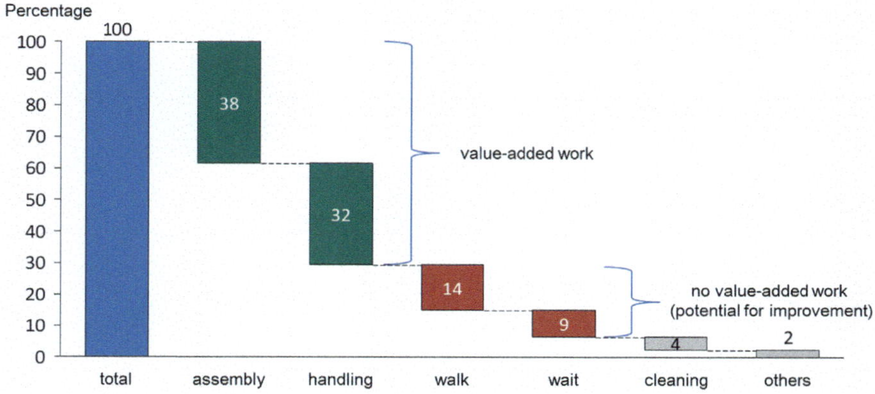

Fig. 8.7 Data collection at the workplace, potential for improvement

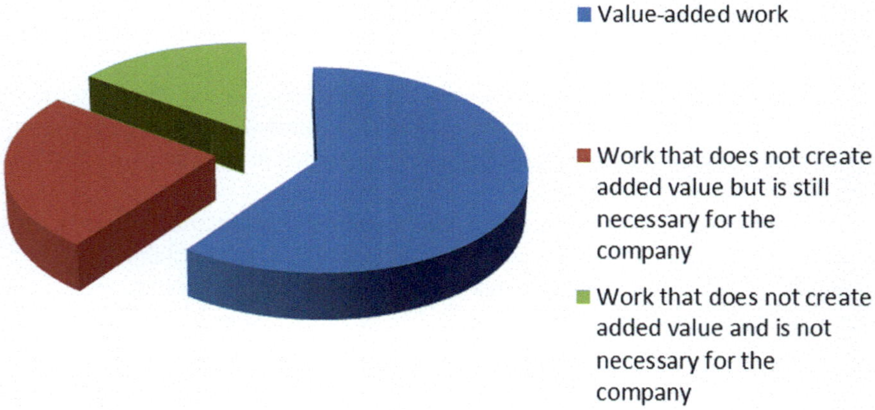

Fig. 8.8 The three types of work in a manufacturing company

2. work that does not add value but is necessary for the company to function properly
3. work that does not add value and is not related to the normal operation of the business

The third category of work is wasteful and must be eliminated.

According to the above categories, there are seven main types of waste.

- Overproduction: producing too many semi-finished and finished products that are not needed
- Excessive inventory: takes up storage space, extends operating time and increases costs
- Too long routes: complex production lines, wasted working hours, waiting at workstations
- Unnecessary transportation: can lead to damage to components
- Waiting times: Equipment breakdowns; raw materials, semi-finished products not available on time
- Production capacity: excessive production capacity, unnecessary automation.
- Rejects and rework: failure to fulfill customer orders on time, damaged parts, rework.

How can waste be avoided?

First analyze the causes of waste, make them transparent and itemized and create a waste list. Then find improvement and optimization measures in the waste list.

8.2.4 5S (5S for a Good Workplace)

A good workplace not only ensures that the manufacturing process runs smoothly, but also reduces the risk of errors and reduces rejects and rework (see Fig. 8.9).

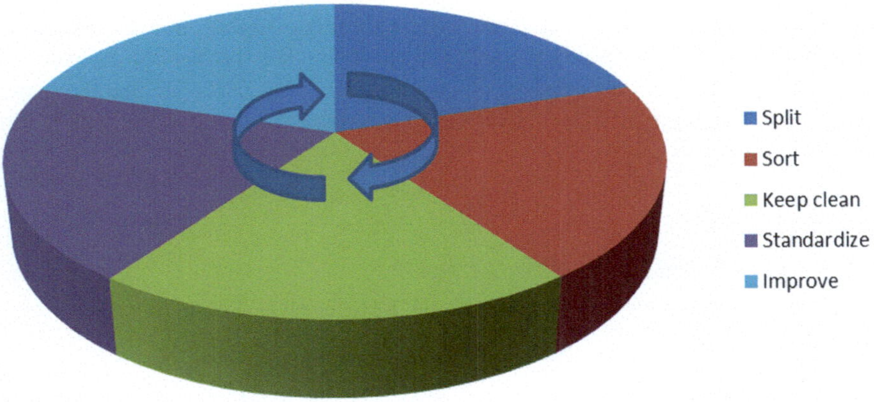

Fig. 8.9 5A and continuous improvement

The first S: Split up

What is needed in the workplace? What is not needed? How often is it needed? Why is it here? Who is responsible for it? For example, tools and documents that are not used.

The second S: Sorting

Are the required tools and accessories in the right place? Are there markings for the areas where tools and accessories are placed, standardization and organization of folders, visualization of operating instructions?

The third S: Keep it clean

Always clean tools and accessories, tidy workstations and common areas.

The fourth S: Standardize

Make these points the norm and the standard.

The fifth S: Improve

All these points are always observed and continuously improved.

Here, the author relates an experience with the optimization of manufacturing costs at Mercedes-Benz. In the final assembly of a passenger car, several hundred screws have to be tightened, large and small, long and short, in different sizes and with different surface coatings. For this reason, the engineers have optimized the tightening process.

Before the optimization:

The normal screw tightening process: the right hand takes the screw, positions it in the hole, then the right hand reaches for the electric screwdriver and tightens the screw.

After optimization:

The head of the screw has been geometrically modified so that the right hand can position the screw in the hole and the left hand then takes the electric screwdriver and tightens the screw. This splits the process of positioning and tightening the screw between the right and left hand, saving 0.5–1 s per tightening process and reducing manufacturing costs.

8.2.5 3M (Avoid 3M in Production)

The first M: "Overload"

z. e.g. overloaded employees, resulting in missing work; overloaded equipment, lack of maintenance time, resulting in expensive repairs and long downtimes.

The second M: "inconsistency"

Different output for the same period; inconsistency between customer orders and production schedules; different skill levels of employees, loss of skilled labor.

The third M: "waste"

Production in excess of customer orders; waiting, searching for raw materials and semi-finished products, unnecessary movement, unnecessary transportation of materials, unnecessary work steps; unused equipment; excessive inventory; errors, rework, scrap.

Lean production is without the three M's (without overload, inconsistency, waste), smooth, error-free, cost-optimized production process.

Chapter 9
Modular Strategy: Cost Engineering 2.0

In 2008, the automotive industry learned a painful lesson after the global economic crisis, and the major companies looked for ways to free themselves from their predicament and develop sustainably.

In 2009, Mercedes-Benz was one of the first manufacturers to propose a modular strategy and defined the Mercedes-Benz front-wheel drive architecture (MFA) and the Mercedes-Benz rear-wheel drive architecture (MRA). The main objective is to reduce costs by 20% (€3 billion) with a total annual procurement volume of €15 billion.

Volkswagen is also proposing a platform strategy with the Modular Transverse Toolkit (MQB) and the Modular Longitudinal Toolkit (MLB). BMW has presented its front-wheel drive (FWD) and rear-wheel drive (RWD) vehicle architectures. In 2015, Toyota announced the Toyota New Global Architecture (TNGA) vehicle architecture, which aims to reduce development costs by 25%

9.1 Why Do Manufacturing Companies Need a Modular Strategy?

Why are these major automotive companies proposing modular vehicle architectures? It starts with the relationship between the cost of components and production volume. As described in Sect. 5.8, there is an analytical function between the manufacturing cost of a component and the corresponding production volume.

An increase in production volume leads to a reduction in the unit cost of a component, mainly because the total investment in the production line can be allocated to a larger number of components. If the production volume exceeds the capacity of a production line, this leads to a jump in the analytical function (see Fig. 5.22).

This is the main reason for modularization.

X. Liu, *Cost Engineering Systems for Manufacturing Enterprises*,
https://doi.org/10.1007/978-981-96-5234-1_9

A component that is used not only in one but in several products significantly increases the production volume and thus reduces the cost per component.

The strategic goals of modularization are

- the increased use of common modules (or subsystems) and components, thus reducing the costs of all the company's products
- to simultaneously reduce R&D costs, tooling costs and the costs of production facilities, and
- reducing R&D risks: reducing the number of new modules (or subsystems) to be developed for a single product and thus improving product reliability;
- the effective use of different modules (or subsystems) from different products over the entire life cycle, which quickly adapt to the different needs of the markets and customers and improve the competitiveness of the products and the economic efficiency of the company

Through the modularization of all products in manufacturing companies the era of Cost Engineering 2.0 begins.

9.2 How to Implement Cost Engineering 2.0?

If we now analyze the construction of a building block (see Fig. 9.1), these building blocks have certain shapes, dimensions and materials. The building block has several small cylinders protruding from the surface with the same diameter and height; the back of the building block is hollow; the building block has a certain thickness. The building blocks are connected to each other in a uniform way (the small cylinder of one building block is inserted into the hollow back of another building block).

There is a relatively limited number of building blocks in different shapes and sizes, but they can be used to build numerous houses, bridges, roads, cars, airplanes, spaceships (toys).

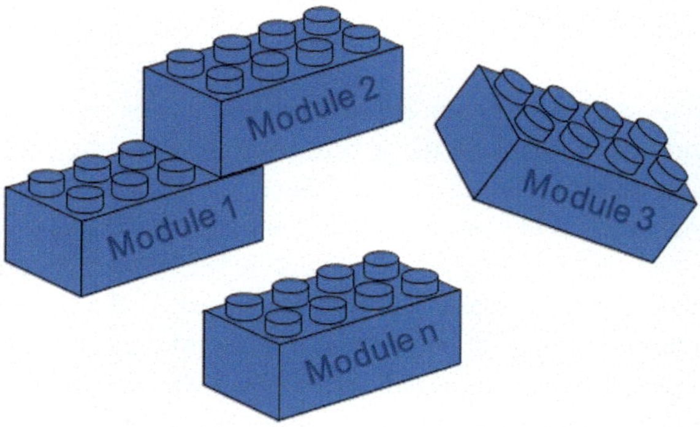

Fig. 9.1 Building blocks and how they are interconnected

The modularization of all products of a manufacturing company can be implemented on the basis of the building block by answering the following questions in particular:

- How many different modules are needed to modularize all (or most) of the company's products?
- What is the technical solution for each module?
- What interfaces are there for the individual modules?
- How are the modules put together to form a product?

The following is an example of the individual modules of a passenger car Table 9.1: The most important criteria for the selection and division of the modules are

- have independent functions
- have defined interfaces (mechanical, fluid dynamic, electrical, etc.)
- be used by the company's many products

Figure 9.2 shows a navigation system module and the associated mechanical and electrical interfaces that meet the requirements of the above splitting criteria.

The strategic vision of modularization is to achieve an optimal level of common use of modules in all the company's products, resulting in

- Standardized processes,
- First-class quality,
- Competitive costs
- Efficient, fast and flexible production systems

Table 9.1 Individual modules of a passenger car

Body in white	Interior and exterior	EE	Chassis	Powertrain
Stem	Interior and exterior mirrors	Management records	Front subframe	Engine
Reinforcing support for the instrument panel	Air conditioning system	Body control modules	Rear subframe	Gearbox
Main bottom	Sunroof	Gateway	Side shafts	Starter
Rear floor	Bumper	Keyless entry	Steering system	Alternator
Side walls	Windshield wipers	Tire pressure control	Brake system	Engine mounts
Doors and flaps	Front seats	Instrument cluster	Rims	Transmission mounts
Mudguards	Airbags, seat belts	Navigation system	Tires	Air intake and cooling system
NVH damping, insulation	Font seats	Telematics box	Pedal system	Exhaust after treatment
……	……	……	……	……

Fig. 9.2 Navigation system and its interfaces (electronic and mechanic)

- Product innovation and performance that customers can experience

This enables the company to develop sustainably in a highly competitive market. The modular strategy follows a clear, predetermined vision, for the realization of which several prerequisites must be created:

Pre-Planning:

Early start of pre-planning to ensure goals are achieved; less pressure for later product development and optimization.

Cross-Departmental Collaboration:

Joint development and cross-departmental coordination to achieve development maturity; full consideration of customer needs, production efficiency and product quality.

Modular Strategy Framework:

The modular strategy is a long-term framework for business activities, not a temporary product project; it requires continuous medium to long-term optimization.

Modular Teams:

Module-based module teams are formed, with core members from the Research and Development, Quality and Cost Engineering departments; all products in the company are planned, designed and implemented by the relevant module teams.

The modular strategy is divided into three phases (see Fig. 9.3).

- Analysis phase (~3 months)
- Technical design phase (~6 months)
- Implementation phase (~24 months)

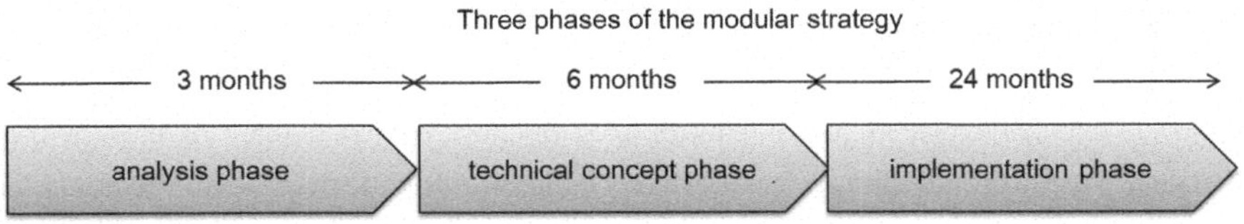

Fig. 9.3 The three phases of the modularization strategy

9.2.1 Analysis Phase

The first starting point is the technical structure of the module.

Which parts (common parts) can be used together for the different products? Which ones need to be differentiated? (Fig. 9.4).

The second starting point is to carefully analyze and optimize the technical design for the module in terms of functional/performance requirements, cost, quality and customer feedback, production time, weight, etc.

Finally, the technical design of the module must undergo a comprehensive evaluation based on the supplier market.

The result of the analysis phase is the current status and future development direction of the module, which will be used by as many of the company's products as possible.

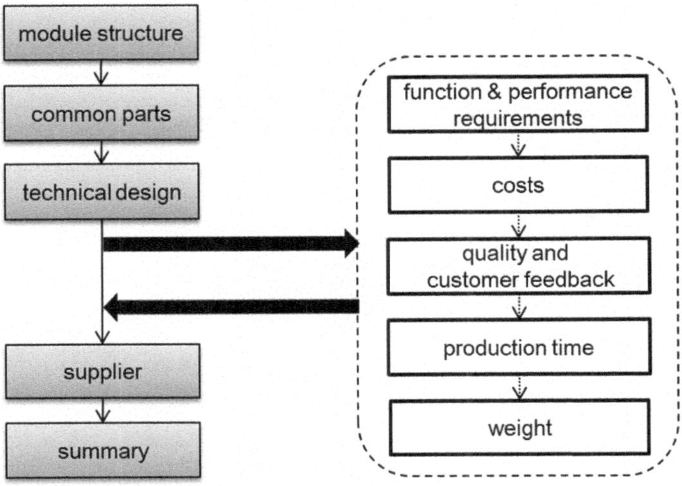

Fig. 9.4 Main working elements of the modularization strategy

9.2.2 Technical Design Phase

In the technical design phase, the module is designed on the basis of the current status and the future development directions of the module derived from the analysis phase using the following three aspects:

1. The division of the module into sub-modules

The instrument cluster module is taken as an example here, which is divided into sub-modules as follows (see Fig. 9.5).

The new technical design leads to the following results:

- The sub-modules stepper motor, circuit board (including electronic components) and pointer are standardized from the original three variants to one

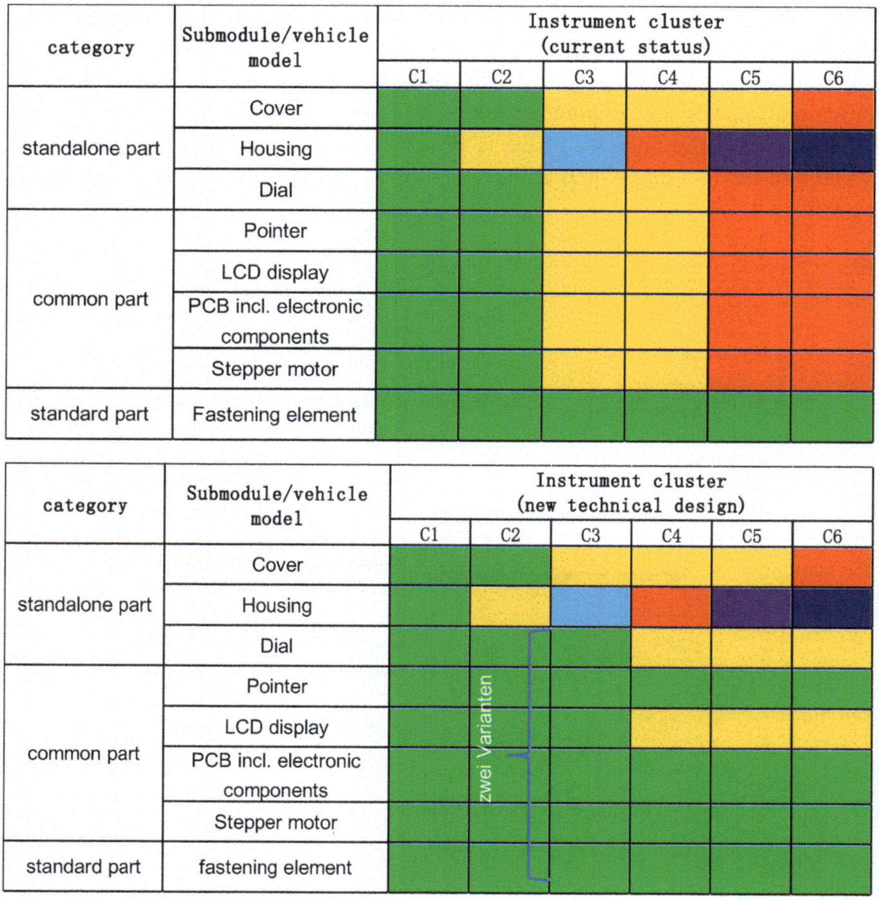

Fig. 9.5 Submodules of the instrument cluster module; [Status quo (top) compared with the new technical design (below)], (The same submodule in different colors means different variants)

- The LCD display and the dial have also been reduced from three to two
- The modular design of the instrument cluster now has a total of two module variants (green and orange colors) in common parts
- The cost analysis

Based on the sales volume of the Common Parts of the modular new instrument cluster module, the cost reduction value of each sub-module was derived using the analytical function of cost versus volume introduced in Chap. 5. 5–10% cost optimization was achieved for the LCD, PCB and stepper motor, and the cost optimization result of the instrument cluster module was reduced from the original 651 CNY to 596 CNY, a cost reduction of 8.4% (Fig. 9.6).

3. Life cycle planning

The life cycle of a module is usually staggered with the life cycle of products, and the life cycle planning of a module should cover two product life cycles (Figs. 9.7 and 9.8).

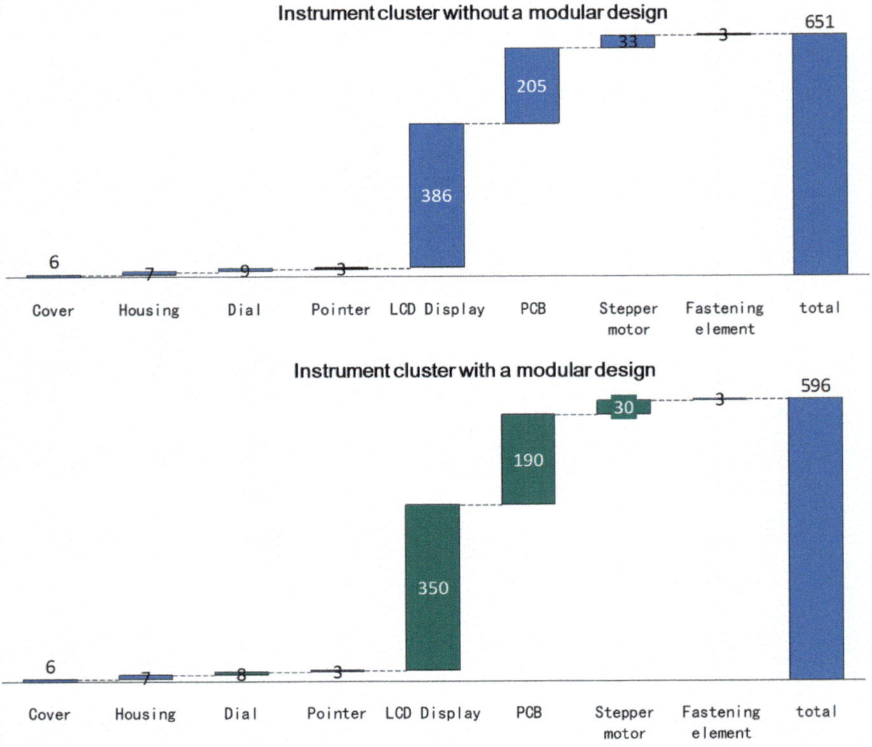

Fig. 9.6 Material cost comparison for the instrument cluster module

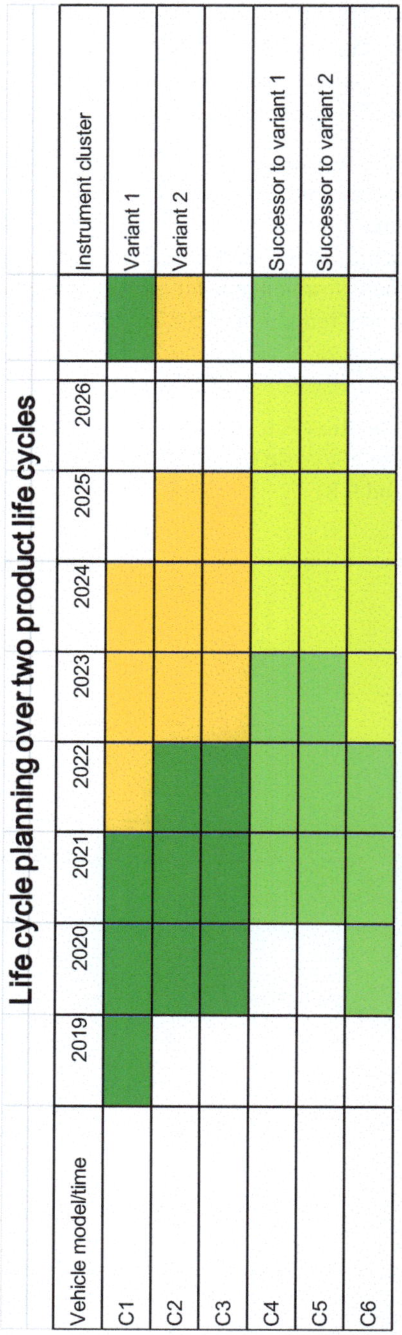

Fig. 9.7 Life cycle planning for the modular instrument cluster module over two product life cycles

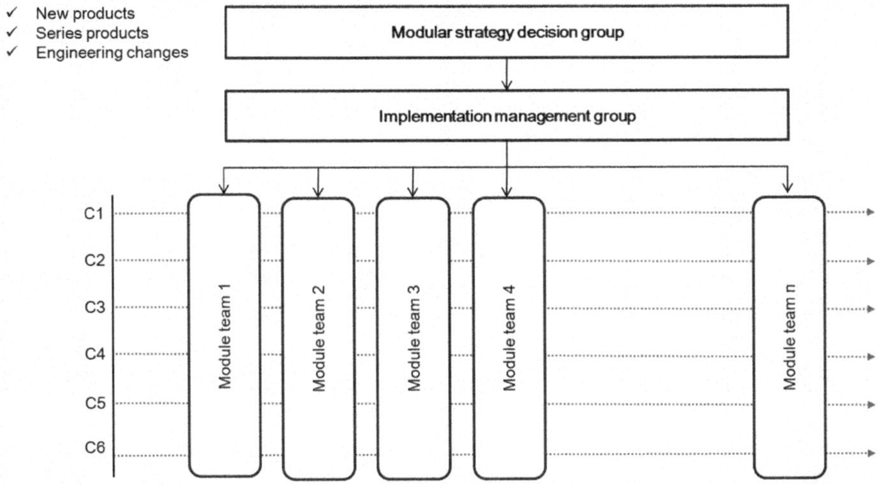

Fig. 9.8 Organizational structure for the implementation of the modular strategy

9.2.3 *Implementation Phase*

A core element of the implementation of a modular strategy is that the company establishes an organizational structure with

- a modular strategy decision group,
- an implementation management group and
- the module teams

All new products, series products and design changes are handled via this structure.

Chapter 10
Cost Engineering for New Energy Vehicles

10.1 The Key Customer Benefits of New Energy Vehicles

What are the most important customer values of New Energy Vehicles?

Safety:

In terms of passive safety, most vehicles can achieve 5 stars from C-NCAP, some even 5 stars plus. ESP is now standard, and the introduction of millimeter wave radar and the use of other driver assistance systems as standard have significantly improved the active safety of the vehicles. A key subsystem of New Energy Vehicles is the battery, whose safety and reliability is crucial.

Range:

As a means of transportation, range is the most important concern of customers or potential customers of New Energy Vehicles. Gradually, car manufacturers have started to label their vehicles with a range of 400 or 500 km on their official websites, instead of acceleration data.

Suitability for everyday use:

Size of the interior and trunk, easy accessibility and operability, everyday usability is one of the most important values for customers.

Comfort:

Seats, ventilation and air conditioning, ride comfort, interior noise and vibration are among the most important values that customers pay particular attention to when using a vehicle.

Affordability:

What most customers expect from new energy vehicles is a safe, practical and comfortable vehicle with sufficient range at a moderate price.

X. Liu, *Cost Engineering Systems for Manufacturing Enterprises*,
https://doi.org/10.1007/978-981-96-5234-1_10

In addition to the body, the interior and exterior and the chassis, the battery, the high-voltage wiring harness, the electric motor and the inverter are the main cost drivers in a new energy vehicle. The precise definition of the vehicle architecture, the functions services required by the customer and the equipment is one of the main tasks in the early phases of the development of a new energy vehicle.

10.2 Cost Engineering in Practice for New Energy Vehicles

10.2.1 Business Case Calculation (NPV) and Sensitivity Analysis

Using the example of an all-electric SUV (see Fig. 10.1), the NPV is calculated over its entire life cycle based on the annual distribution of the forecast sales volume and the forecast sales price of CNY 391 million (blue curve).

Sensitivity analysis:

If the projected sales volume decreases by 5%, the NPV will drop to CNY 362 million (green curve).

If the forecast sales price is reduced by 5%, the NPV will fall to CNY 138 million (curve in purple color).

If the target costs exceed 5%, the NPV decreases to CNY 193 million (red curve).

The business case calculation shows that if the costs exceed the target costs by 5%, the NPV is almost halved. The high material costs (battery pack, inverter, electric motor, high-voltage wiring harness, etc.), the enormous investments in research and development and the overly optimistic market forecasts (sales volume and sales price) lead to a "survival-threatening" situation for an automotive company.

The only way to succeed in a competitive market with many competitors in the future is to have an accurate product definition and market forecast (sales and price), to design the vehicle at the target cost and to perform an accurate business case calculation and its sensitivity analysis.

10.2.2 Calculation of the Reference Cost of Lithium Battery Cells

In 2021, about 3.3 million new energy vehicles with a total installed battery capacity of about 150 GWh were produced in China. Good planning of the total vehicle costs for new energy vehicles and the optimization of battery costs are the core tasks of vehicle manufacturers and system suppliers for the coming years. Currently, lithium batteries are mainly used in new energy vehicles. There are two main types of active materials for the positive electrode: lithium–nickel–cobalt–manganese compounds

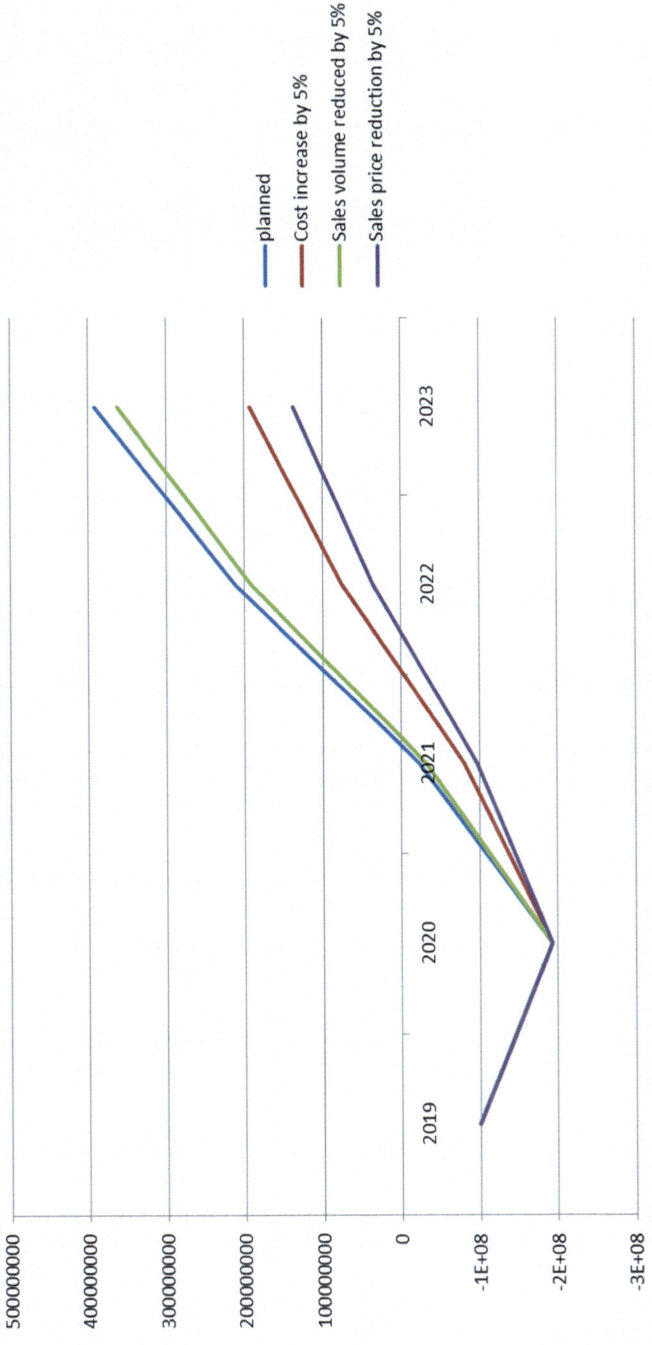

Fig. 10.1 Business case calculation and sensitivity analysis for an electric SUV

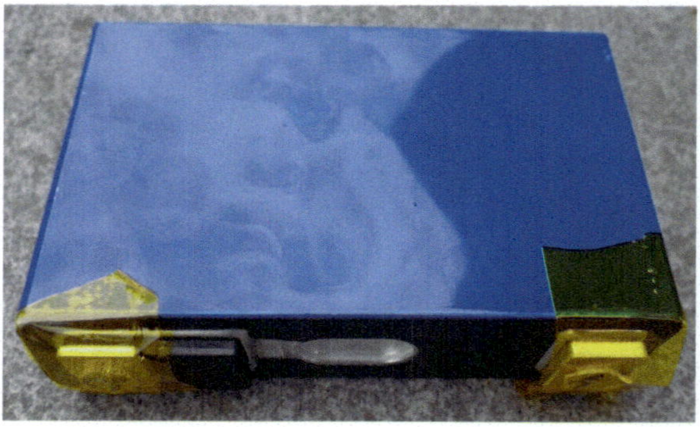

Fig. 10.2 Prototype of a Li-ion battery cell (NCM111, 255 Wh, 3.65 V)

(NCM) and lithium–iron–phosphate (LFP). The negative electrode consists mainly of graphite (95%) and a few additives. A battery cell is the smallest unit of the battery pack, and the cell module consists of several interconnected cells, and the battery pack consists of several cell modules and the battery management system, etc. Accurately calculating the cost of battery cells, identifying the main cost drivers, and then optimizing the technology is an important way to optimize the cost of the battery pack and thus the cost of the new energy vehicle.

1. Disassembly of the lithium battery cell

Only the most important steps in dismantling the battery cell are described here. The disassembly of lithium battery cells must be carried out in a professional laboratory, which is a guarantee of operational safety.

First, the lithium battery cell is completely discharged over a longer period of time. The lithium battery cell used for disassembly in this example is a prismatic cell with a voltage of 3.65 V, a capacity of 255 Wh and a gross weight of 1.4 kg (see Fig. 10.2).

Subsequently, the cell case was opened in the dry laboratory at low temperature, the layer thickness and the outer dimensions of the positive film or NCM material, the negative film or graphite were measured (see Figs. 10.3, 10.4 and 10.5), where the positive film has an active compound (NCM) and an aluminum substrate, and the negative film contains an active compound (graphite) and a copper substrate, thus obtaining the main materials for the cost calculation of lithium battery cells.

2. Material composition of the lithium battery cell

The material costs of this lithium battery cell were calculated using the cost engineering SaaS software CostKey [3] (Fig. 10.6).

The most important cost factors are NCM material, electrolyte, copper foil, etc. The optimization of material costs is based on these cost drivers, such as NCM material (NCM), where C refers to the expensive cobalt material. Reducing the cobalt content is the primary option for optimizing material costs. Switching from NCM111 to NCM523 or even NCM811 means reducing the cobalt content while increasing the nickel content. Of course, increasing the nickel content also affects

Fig. 10.3 NCM layer of a Li-ion battery cell

Fig. 10.4 Graphite layer of a Li-ion battery cell

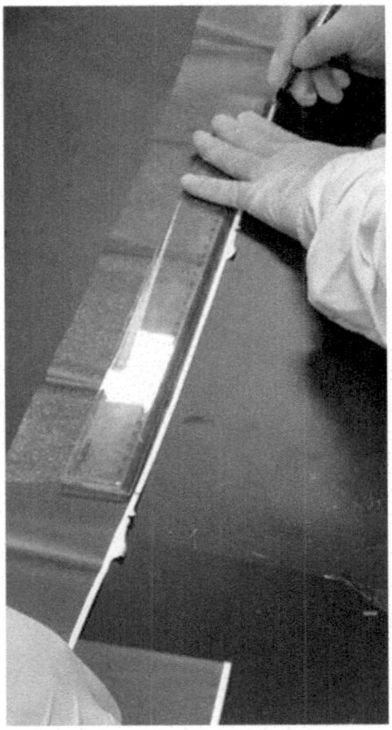

Fig. 10.5 Disassembly of a lithium battery cell and measurement in a professional laboratory

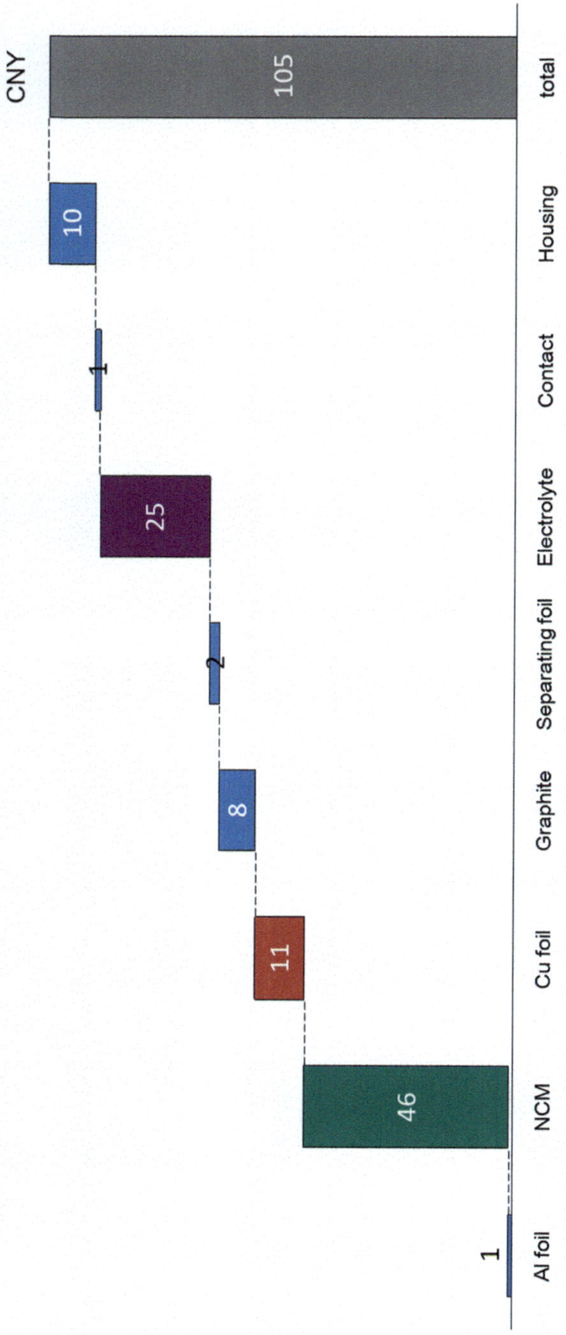

Fig. 10.6 Breakdown of the material costs of the lithium battery cell

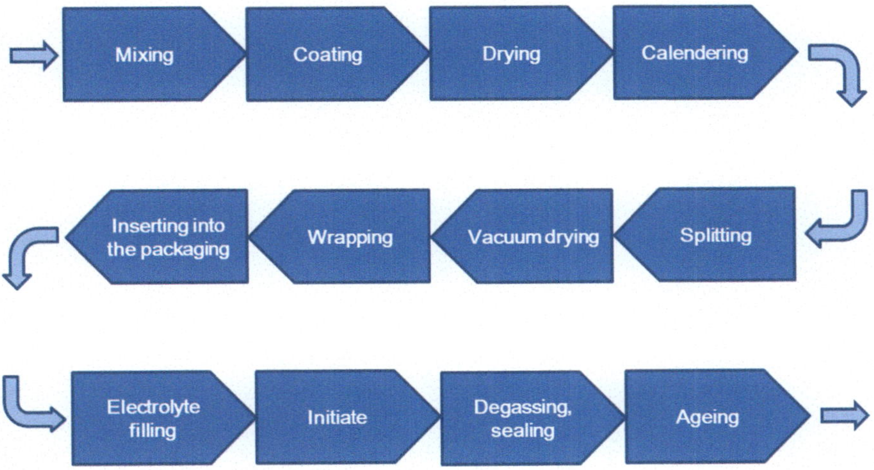

Fig. 10.7 Lithium battery cell manufacturing process

the stability of the cell, so a balance must be found between stability and material costs in order to achieve optimization.

3. Manufacturing process of the lithium battery cell

The production of lithium battery cells can be divided into about 12 processes (see Fig. 10.7).

The processes of mixing, coating, drying, calendering, vacuum drying, splitting and ageing are batch processes, while the processes of stacking/winding, electrolyte filling and sealing etc. are clocked processes.

The manufacturing costs of this lithium battery cell can also be calculated using the cost engineering SaaS software CostKey [3].

The results of the manufacturing cost calculation (see Fig. 10.8) show that the manufacturing costs are largely determined by the clocked process steps such as stacking/wrapping, electrolyte filling, sealing, etc.

Figure 10.9 shows the material costs, manufacturing costs and overhead as well as the total reference costs.

10.2.3 Comparison of the Structure and Costs of Directly and Indirectly Driven Electric Motors

The drive unit of a pure electric vehicle usually consists of an inverter, a permanent magnet AC synchronous motor (PMAC synchronous motor) and a reduction gearbox. In order to optimize the design, a comparison is made below between alternatives with direct and indirect drive.

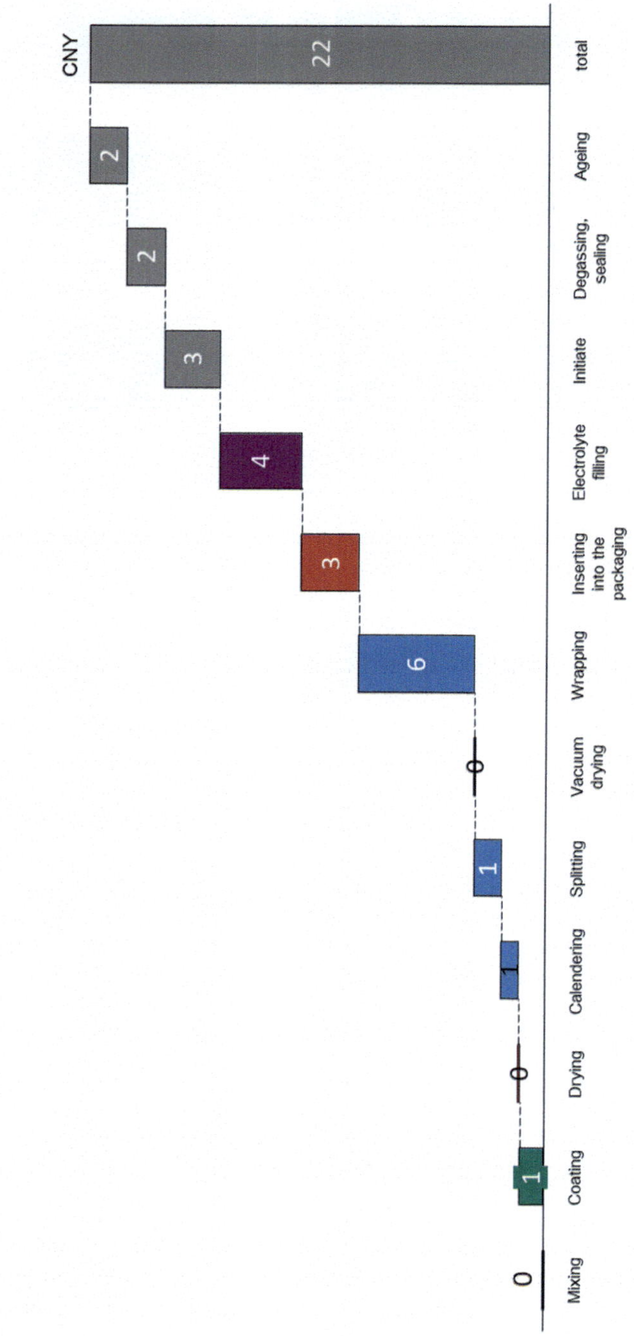

Fig. 10.8 Breakdown of the manufacturing costs of a lithium battery cell

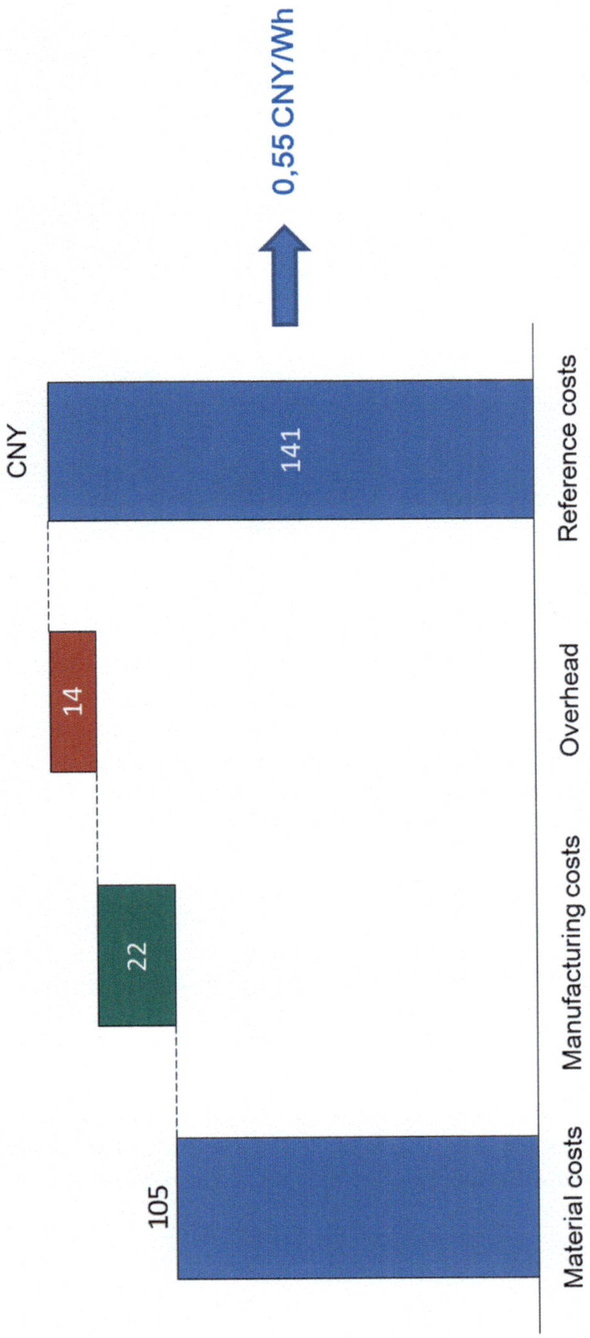

Fig. 10.9 Reference costs of a lithium battery cell and their distribution

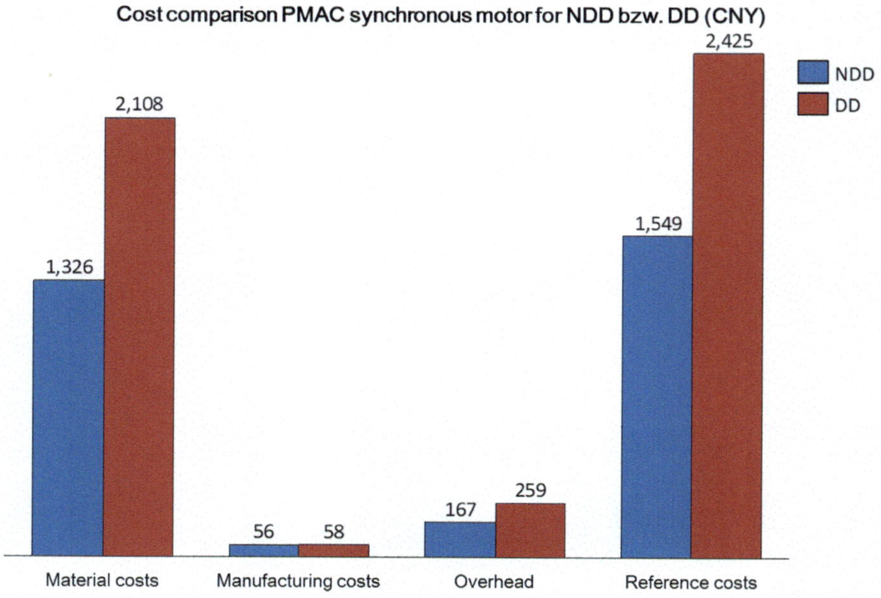

Fig. 10.10 Comparison of the reference costs of PMAC synchronous motor for NDD and DD

In the indirect drive module (non-direct drive: NDD), a PMAC synchronous motor and a reduction gearbox form the overall package, while the direct drive module (direct drive: DD) consists only of a PMAC synchronous motor without a reduction gearbox; both electric motors have a maximum output power of 60 kW. After completing the electric motor designs, the reference costs of the PMAC synchronous motor NDD and DD were calculated (see Fig. 10.10) and compared as follows.

The reference cost of the NDD motor is 1549 CNY, and together with the cost of the gearbox (about 900 CNY), the total cost of the indirect drive module is about 2449 CNY, while the total cost of the direct drive module (DD) is only 2425 CNY (see Fig. 10.11).

The weight of the NDD motor is 31.9 kg, that of the reduction gearbox ~16 kg, i.e. a total of ~47.9 kg; the DD motor weighs 55.3 kg.

To summarize:

The direct drive module has NVH advantage, and more cost-effective; the indirect drive module is lighter (weight advantage). The optimal choice depends on the positioning and the functional/performance requirements of the product.

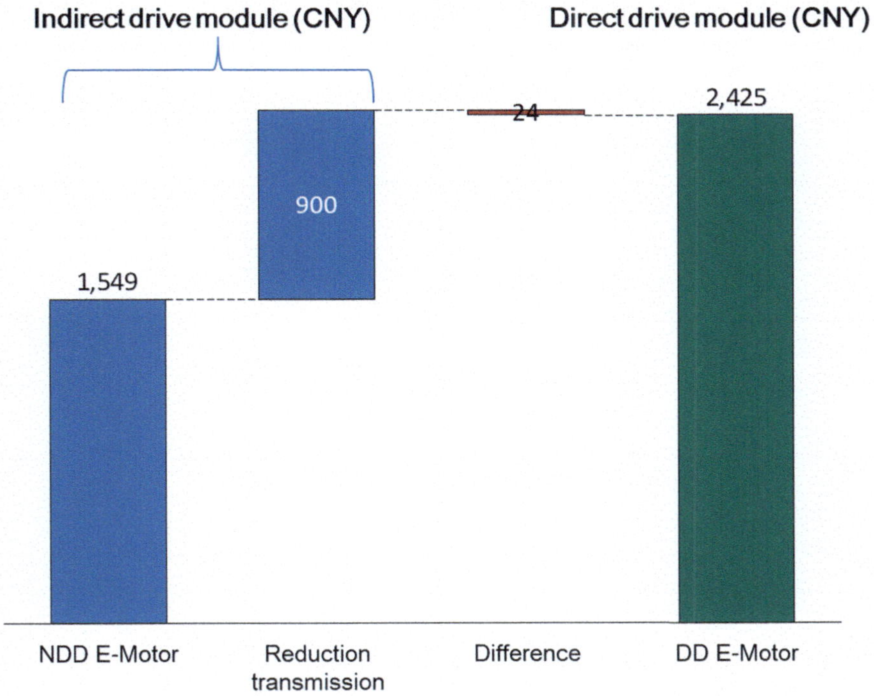

Fig. 10.11 Reference cost comparison of indirect drive module and direct drive module

Chapter 11
Digitalization of Cost Engineering

Digitization is everywhere, if you first look at the products and services:

- Satellite navigation systems based on digitized, high-resolution maps, and positioning signals from satellites
- Numerous app applications in cell phones: e.g. booking a trip, ordering a cab, ordering food
- Remote control of household appliances in smart homes (refrigerators, washing machines, surveillance cameras, etc.)
- ID card, driver's license stored in digital form in cell phone; digital currency, payment via QR code
- Cloud: photos and files stored in the cloud, accessible anywhere; software as a service (SaaS) can be used anywhere on the cloud without local installation; online meetings worldwide

If you now look at production:

- CNC machine tools, industrial robots and automated production lines
- Cloud-based manufacturing execution system (MES)
- Intelligent material flow and warehouse management system
- In-line software flash
- Internet of Things

The list will be very long. Digitalization, networking and intelligence of products, services and production simultaneously require the digitalization of cost engineering.

11.1 Digitalization of the Product Process

Design:

X. Liu, *Cost Engineering Systems for Manufacturing Enterprises*,
https://doi.org/10.1007/978-981-96-5234-1_11

The digitalization of the product process is reflected in the decades-long shift from drawing components on paper to using CAD software systems for designing and drawing components; they are represented by CATIA, Pro-Engineer, Uni-Graphics, etc. Designers use computer-aided styling Computer Aided Styling (CAS). In the design of components, finite element calculations are used extensively for the static and dynamic analysis of strengths and deformations, as well as for flows, thermal and electromagnetic fields.

Production:

CNC machines, robots, software-controlled production lines are widely used in practice.

Planning and management:

Software systems for customer orders, production planning, material purchasing and logistics, customer service have become commonplace in business.

So what are the applications and innovations in cost engineering in the digitalization of the product process?

11.2 CAE for Optimizing Product Material Costs

Computer Aided Engineering (CAE) is used extensively in product development, with finite element calculations being the most widely used.

Here is an example: through finite element calculations of the static and dynamic torsional and bending stiffness of the entire vehicle, the conclusions drawn can optimize the design to reduce or eliminate the use of more expensive structural adhesives in some places (see Fig. 11.1), and some of the structural adhesives can be replaced with low-cost sealant, reducing the cost of a vehicle by about 53 CNY (Table 11.1).

Another example is the weight and cost optimization of a wheel rim (see Fig. 11.1).

The optimization leads to a solution that reduces the weight of the rim by approx. 250 g and lowers the costs by approx. 8 CNY.

11.3 CAE Optimization of Manufacturing Costs

Simulation calculations can not only help to optimize product costs, but also help manufacturing process engineers to optimize manufacturing processes and reduce manufacturing costs.

Figure 11.2 shows a calculation result of the paint layer thickness distribution in a painting line with a painting robot, which moves along the programmed paths and spray paints the body shell of a passenger car. The programming and simulation system was significantly co-developed by the author at the Fraunhofer Institute for Production Systems and Design Technology in Berlin and later industrialized as

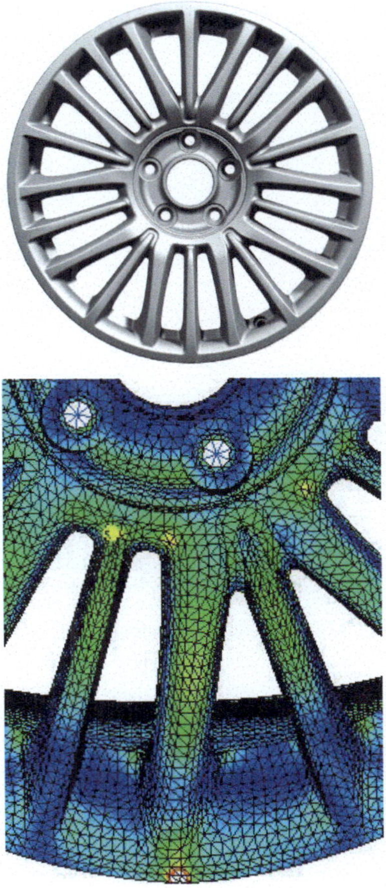

Fig. 11.1 Finite element calculations of a rim when approaching the edge of the sidewalk at 13°

Table 11.1 Sealant and structural adhesive for a passenger car

Measure	Material	Length (m)	Costs (CNY)	Cost saving (CNY)
Sealing compound replaces the structural adhesive	Structural adhesive (EPA 240 H)	1,4	45	23
	Sealing compound (IM1020)	1.4	22	
Omission	Structural adhesive (EPA 240H)	1.0	0	30

the RobCAD Paint Module software module from Siemens PLM. Some well-known automotive companies use the Paint Module for robot programming and optimization of the painting process.

In the virtual painting process, costs can be reduced by varying the painting process parameters (paint flow rate, spray cone geometry, painting distance) and

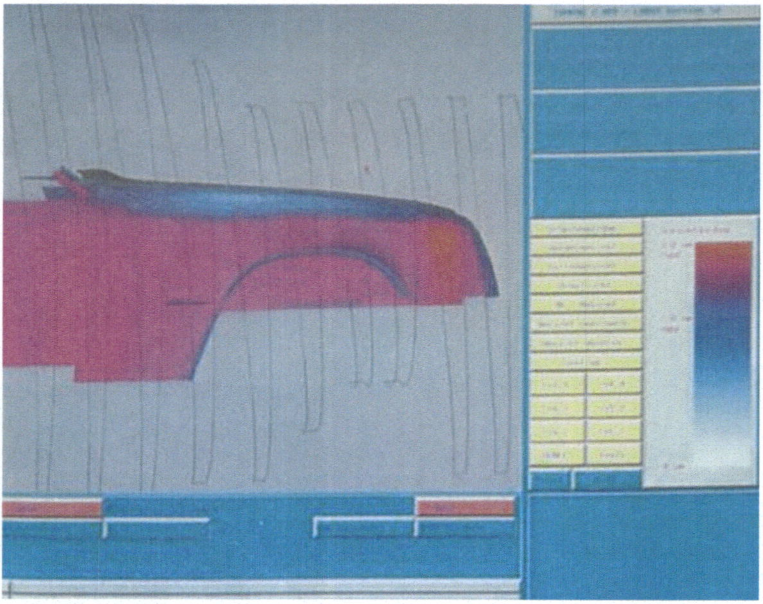

Fig. 11.2 Programming and process simulation system for painting robots

by optimizing the robot path geometry and speed to reduce overspray and optimize the painting time. In addition, the quality of the paint layer can be guaranteed by programming the robot paths in such a way that insufficient (too low coverage) or excessive paint layer thickness (tends to run) can be avoided.

11.4 Cost Database

A cost database can be used to

- Query raw material costs, production equipment costs, labor costs, etc.
- Carry out a competitive analysis
- To plan R&D costs, tooling costs, investments in production lines

An example of the use of the cost database in CostKey [3].

If one wants to query the raw material cost per kilogram of PA66 GF30, enter PA66GF30 in the search window of CostKey, the result is 28 CNY per kg (see also Fig. 11.3).

If you want to know the investment amount of an injection molding machine, enter it in the search window and you will get the query result from the cost database, e.g. the investment amount of a 2000 tons injection molding machine is 9.9 million CNY. This cost data can be used to calculate manufacturing costs or as a basis for planning production facilities.

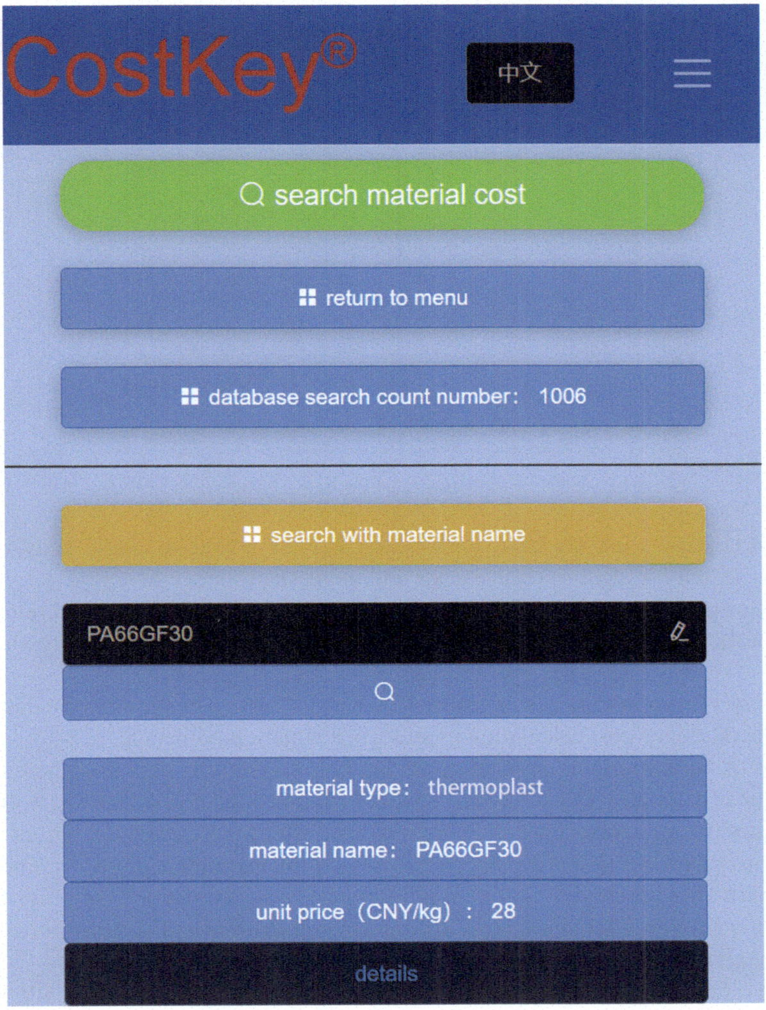

Fig. 11.3 Querying cost data in CostKey [3]

11.5 Cost Software

Cost software is a necessary tool to efficiently and effectively implement "Design to cost", "Value Analysis & Value Engineering (VAVE)", and in general "Cost optimization". In manufacturing companies, the following departments or engineers can use cost software on a daily basis.

Design engineers can query basic cost data as well as the reference costs of competitive products calculated by cost engineers, and thus support them in designing to target costs

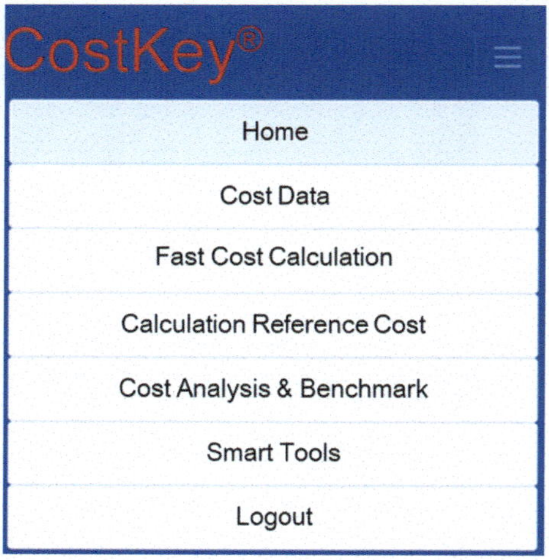

Fig. 11.4 Cost software CostKey [3] as SaaS (Software as a Service)

Cost engineers use the cost software to calculate reference costs, analyze cost drivers, search for ideas to reduce costs and increase product efficiency.

Buyers can check supplier offers and compare them with the calculated reference costs of the cost engineers to provide technical arguments for business negotiations etc.

Figure 11.4 shows the cost software CostKey [3], which contains a cost database module, reference cost calculation module, benchmark module, etc.

One of the important applications of the cost software is explained here using a practical example. The reference cost calculation module of the cost software calculates the reference costs of an instrument cluster for a car whose current place of origin is the USA. What are the cost savings if the instrument cluster is produced in China? This is called a "differential reference cost calculation" (see Fig. 11.5), where the difference is mainly in the different manufacturing and administrative costs (overhead) in China and in the USA.

The reference costs for the combination instrument are around 356 CNY, but if it is manufactured in China in the future, the reference costs will fall to 280 CNY, which corresponds to a cost optimization of 21%. The same approach can be used to analyze the cost difference between the production of the electric vehicle Tesla Model 3 in the USA and the production in China (Shanghai).

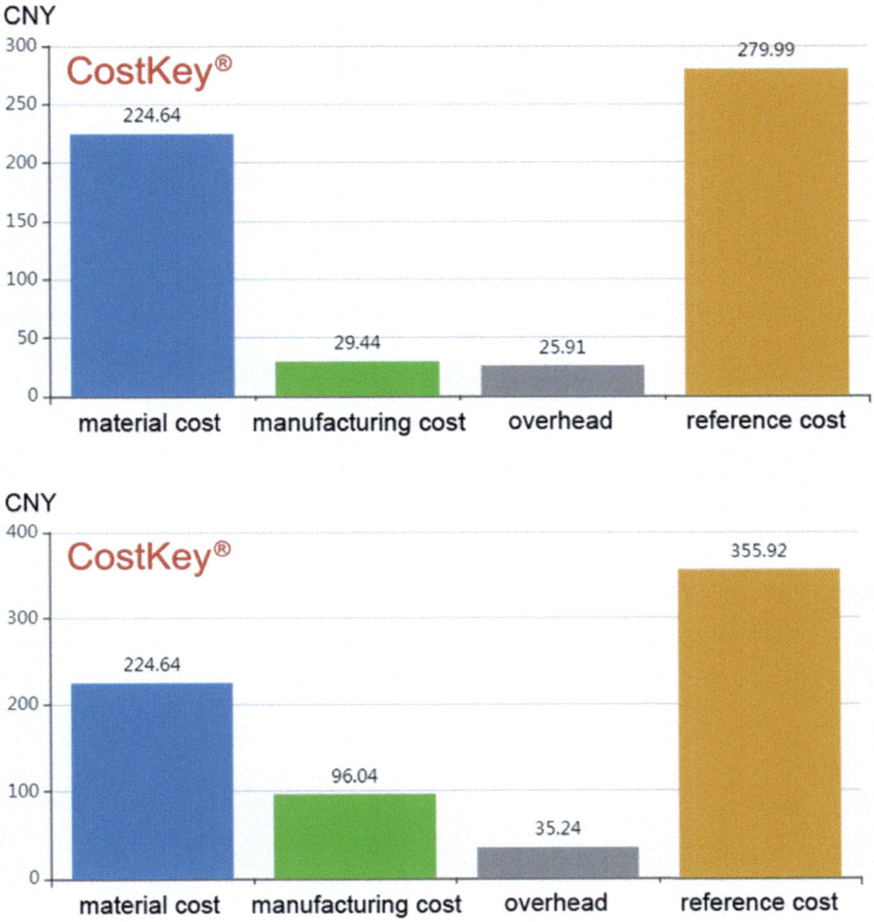

Fig. 11.5 Reference cost calculation of a combined instrument using CostKey [3] top origin China, below origin USA

Conclusion

Based on 25 years of practical experience in product and process development, as well as cost engineering in manufacturing companies (Mercedes-Benz and Great Wall Motor), the author has systematically introduced cost engineering system, cost engineering methods and processes for the first time in this book and shown many practical examples of cost engineering.

Every manufacturing company needs a fully established and efficiently functioning cost engineering system, including the independent positioning of cost engineering within the company organization, the cost engineering methods and processes embedded in the product process, and a clearly defined performance assessment mechanism (KPI) that is actually implemented and valid for all employees of the company.

Cost engineering is a core function of the company, just like R&D and quality.

The cost engineering system must be standardized. If a company is certified according to the cost engineering system standard (ISO/GB), this is, like the certification of the quality system (ISO 9001), a "brand mark" that the company's processes and efficiency in the area of cost engineering are at a certain high level.

The development of a standard for cost engineering systems is still in its infancy and requires benevolent support and careful detailed work from standardization institutions, research institutes, production companies, etc.

With the digitalization and intelligence of products, cost engineering will also enter the digital age in the future. Cost databases, cost software (SaaS) and (Artificial Intelligence) AI-supported design to target costs will be the future direction of cost engineering.

X. Liu, *Cost Engineering Systems for Manufacturing Enterprises*,
https://doi.org/10.1007/978-981-96-5234-1

References

1. Cost engineering system in the automotive industry - key to reducing costs and increasing efficiency increase, Dr. Xiaoyi Liu, published on 14.08.2017 in Gasgoo-Auto
2. Techniques of Value Analysis and Engineering, Third Edition, 1989, Lawrence D. Miles
3. CostKey is a trademark registered in "China National Intellectual Property Administration". Trademark of CostKey Solutions (Shanghai) Co. Ltd, and is also a SaaS (Software as a Service) SaaS (Software as a Service) for cost engineering. CostKey Solutions (Shanghai) Co. Ltd. has all copyrights of CostKey as SaaS, see also the certificate No. 7197150 issued by National Copyright Administration of the People's Republic of China

X. Liu, *Cost Engineering Systems for Manufacturing Enterprises*,
https://doi.org/10.1007/978-981-96-5234-1

Index

A
American manufacturing companies, 20
Analysis category, 45

B
Benchmark, 27, 34, 59, 71
Benchmarking, 27
Best of bench, 71
BMW, 19, 30, 40, 66, 133
Bosch, 19
Business case calculation, 5, 6, 13, 22, 45, 82, 83, 91, 108, 113, 114, 120, 144, 145

C
Chang'an, 21
Chinese manufacturing companies, 21, 22
Commercial category, 45
Cost database, 158, 163
Cost engineer, 4
Cost engineering 2.0, 133, 134
Cost engineering embedded in the product process, 18
Cost engineering methods, 2, 4–6, 12, 13, 16, 29, 36, 45, 85, 163
Cost engineering system, 2, 3, 5, 12, 13, 15–17, 22, 163
Cost engineering system standard, 12
Cost estimators, 17, 19, 20
Cost software, 159, 160
Cost-weight function, 102
Customer complaints, 31, 41
Customer satisfaction survey, 31

D
Daimler, 15, 17, 38, 74, 93, 107
Design category, 45
Design to cost, 5, 6, 22, 27, 53, 108, 159
Development schedule, 8
Digitalization of cost engineering, 155
Door panel, 56, 57, 64, 71, 85
Dornier Luftfahrt, 107

E
Electronic airbag control unit, 53, 54
Engineering changes, 1, 5, 20, 22, 110, 113
Entire development process, 37
European manufacturing companies, 17

F
First customer, 111
5S, 129
Ford, 20
Forecast sales price, 22, 107, 108, 113, 144
Forecast sales volume, 22, 82, 108, 144
Fuel tank, 69, 115
Functional/performance, 4
Function carriers, 25, 27, 72
Function-cost matrix, 56, 58
Functions, 19, 23–25, 27–32, 34, 38, 40, 43, 56, 59, 62, 63, 66, 68, 93, 108, 117, 135, 144

G
Geely, 21
General electric, 23
General motors, 20
Great Wall Motor, 21, 86, 110, 163

X. Liu, *Cost Engineering Systems for Manufacturing Enterprises*,
https://doi.org/10.1007/978-981-96-5234-1

H

History of cost engineering, 16

I

Information flow, 123
Initial-BoM, 22
Instrument cluster module, 138–140

J

Japanese manufacturing companies, 19

K

KPI, 2–4, 12, 13, 163

L

Lawrence D. Miles, 23
Life cycle management, 27
Life cycle planning, 139, 140
Lightweight construction, 101–103
Linear regression analysis, 68, 69

M

Manufacturing costs, 38, 47, 49, 51, 52, 65, 72, 75, 92, 101, 121, 122, 126, 130, 149, 150, 156, 158
Matching process, 118
Material costs, 47, 49, 51, 52, 75, 81, 101, 107, 121, 144, 146, 148, 149, 156, 158
Material flow, 123
Mathematical model, 50
Mercedes-Benz, 30, 40, 66, 79, 80, 110, 123, 130, 133, 163
Method of data collection at the workplace, 127, 128
Millimeter-wave radar, 51
Modular strategy, 27, 133, 136
Morphological box, 5, 54, 55, 57
Multi-dimensional ABC analysis, 45

N

Negotiation strategy and tactics, 5, 45, 74
Net present value, 5, 6, 82
New energy vehicles, 143
Normalized depreciation costs, 74

O

One-time costs, 82
Option list, 34, 35
Organizational structure of the cost engineering system, 4
Original price planning, 19
Over-engineering, 93, 98–100
Overhead, 47, 49, 51, 75, 149, 160

P

Passenger car, 7, 9–11, 24, 26, 27, 35, 42, 45, 51, 53, 56, 58, 64, 67, 69, 71, 72, 88, 95, 99, 101, 122, 130, 135, 156, 157
Performance evaluation mechanism (KPI), 4
PMAC synchronous motor, 149, 152
Portfolio analysis method, 71
Process category, 45
Product definition, 6, 19, 29, 41, 42, 113
Product development phase, 5, 6, 36, 43, 68
Product function, 23
Product function/performance, 16, 23, 56
Production volume, 74
Product manufacturing phase, 38
Product process, 4–7, 12, 13, 15–18, 29, 30, 41–43, 108, 156, 163
Product sales and after-sales service phase, 38
Product specification, 32, 34, 87, 93, 95, 98
Product value, 23, 56

Q

Quartz watch, 23–25, 56

R

R&D cost planning, 7, 9, 10
RASIC, 1, 41, 42
Reduce component costs, 85
Reference cost calculation, 5, 6, 13, 27, 47, 161
Reference cost of lithium battery cells, 144
Reliability, 31
Research and development system, 1
Reusable plastic box, 68

S

7V, 127
Simulation system for painting robots, 158
Simultaneous engineering, 27
Star parts, 20
System, 1

T

Tactics of the negotiation process, 76
Target cost achievement path, 117, 119
Target cost process, 24, 45, 107–109, 120
Target costs, 7, 12, 13, 17, 22, 43, 53, 59, 68, 82, 108, 110, 112, 113, 115–117, 144, 159, 163
Target customer analysis, 32, 34
Tasks and responsibilities of cost engineering, 43
Technology path, 32–34, 36
Three key elements of a cost engineering system, 3
3M, 130
Three phases of the modularization strategy, 137
Tool cost planning, 10
Tooling cost break-down, 11
Toyota, 15, 19, 20, 30, 40, 66, 121, 123, 133

V

Value analysis, 6, 27, 61, 68
Value Analysis & Value Engineering (VAVE), 4, 5, 7, 13, 15, 22, 23, 38, 60, 85–87, 98, 108, 110, 159
Value engineering, 5, 6, 15, 23, 60, 85–87, 159
Value management, 23, 27, 28
Value stream mapping, 123, 124

The manufacturer's authorised representative in the EU is Springer Nature Customer Service Centre GmbH, Europaplatz 3, 69115 Heidelberg, Germany. If you have any concerns regarding our products, please contact ProductSafety@springernature.com

Printed and bound by CPI Group (UK) Ltd, Croydon, CR0 4YY
01/10/2026
02225945-0015